TECHNOLOGICAL SUPPORT FOR WORK GROUP COLLABORATION

T0187476

edited by

MARGRETHE H. OLSON
New York University

CRC Press
Taylor & Francis Group
Boca Raton   London   New York

CRC Press is an imprint of the
Taylor & Francis Group, an **informa** business

The line drawing on the cover of this book was done by
Steve Osburn, and is adapted from an illustration
in chapter 8 of this book.

First published 1989 by Lawrence Erlbaum Associates, Inc.

Published 2008 by CRC Press
Taylor & Francis Group
6000 Broken Sound Parkway NW, Suite 300
Boca Raton, FL 33487-2742

© 1989 by Taylor & Francis Group, LLC
CRC Press is an imprint of Taylor & Francis Group, an Informa business

First issued in paperback 2019

No claim to original U.S. Government works

ISBN 13: 978-0-367-45120-2 (pbk)
ISBN 13: 978-0-8058-0304-4 (hbk)

**Visit the Taylor & Francis Web site at
http://www.taylorandfrancis.com**

**and the CRC Press Web site at
http://www.crcpress.com**

Technological support for work group collaboration / edited by
Margrethe H. Olson.
     p. cm.
  Based on contributions to the Symposium on Technological Support
for Work Group Collaboration, held by New York University's Center
for Research on Information Systems, May 1987.
  Bibliography: p.
  Includes indexes. ·
  ISBN 0-8058-0304-1
  1. Work groups--Data processing--Congresses. 2. Information
technology--Congresses. I. Olson, Margrethe H. II. Symposium on
Technological Support for Work Group Collaboration (1987 : New York
University, Graduate School of Business Administration, Center for
Research on Information Systems) III. New York University.
Graduate School of Business Administration. Center for Research on
Information Systems.
HD66.T44 1989
658.4'036'0285--dc19   88-30890
                                           CIP

CONTENTS

PREFACE

What is technological support for work group collaboration? It is a label for an emerging trend in product development as well as a body of research on the relationship between computer systems and social systems. The emphasis in both research and product development is on building tools to help people work together. Part of the research is concerned with assessing the ways in which people use computer-based tools to work together and understanding how the tools and the work relationship affect each other. Some of the research focuses on tools that have been around for a while—electronic mail, computer conferencing, project management, etc. Some of it focuses on brand new tools emphasizing "work group productivity" rather than individual productivity. The underlying common theme is a new way of thinking about work—in particular, knowledge work—and the ways it can be made more effective and personally satisfying through information technology.

The notion of technological support for work group collaboration is based on three underlying assumptions:

- Knowledge work is largely cooperative in nature. According to Webster's New World Dictionary, to cooperate means "to act or work together with another or others for a common purpose." Personal computing primarily supports individual work. The emphasis of most "knowledge work" mechanization or augmentation efforts has been primarily on improving individual productivity, ignoring the cooperative nature of the work performed.
- Connectivity brings the potential to support work groups, but current tools are inadequate. The promise to bring connectivity to personal computing, to be able to share data, to provide integration of data resources across all forms of computing, is real. The environment necessary to support work groups will exist, but current software tools

are inadequate because they emphasize primarily either centralized data control or enhancement of personal productivity.

- We do not understand the nature of cooperative work in intellectual efforts; hence, it is difficult to build tools to support it. The collaboration of computer scientists with social scientists—social psychologists, sociologists, and anthropologists—is essential for this work to progress.

In May 1987, New York University's Center for Research on Information Systems ran a Symposium on Technological Support for Work Group Collaboration. This was the second conference on the subject, the first being the Conference on Computer Support for Cooperative Work held in Austin, Texas in December 1986. The purpose of the NYU Symposium, which is an annual event, is to bring together researchers and practitioners to address an area of research which is beginning to emerge in practice. Thus the symposium emphasized products and practical needs for work group support as well as current research.

The papers in this book are based on the contributions of the authors to the NYU Symposium.

In the first chapter (User Approaches to Computer-Supported Teams) Robert Johansen of the Institute for the Future provides a tour of seventeen different approaches to using computers to support work teams. By showing the range of approaches, Johansen demonstrates that the "field" of technological support for collaboartion is still emerging. He shows how many seemingly unrelated tools, from video conferencing to project management, can be labeled as team support. Johansen makes some cogent predictions of how the "field" will be emergently defined and which will be the key users, vendors, and products in the next five years.

The chapter by Vasant Dhar and Margrethe Olson of New York University (Assumptions Underlying Systems that Support Work Group Collaboration) defines work group collaboration in terms of contract definitions and executions. They argue that most systems that purport to support collaboration have a limited view of that support: either communications (e.g., electronic mail) or problem solving (e.g., project management), but not both. Through a case scenario, they illustrate the limitations of both types of systems and demonstrate the power of an integrated tool that would support both problem solving and communications.

Bonnie Johnson (How is Work Coordinated? Implications for Computer-Based Support) assesses tools for coordination from the perspective of Corporate Technology Planning at Aetna Life and Casualty. She contrasts current approaches to design of coordination systems, based on artifacts of existing methods such as memos and telephone message slips, with an approach to design based on conversation. She gives several examples of current projects at Aetna implementing tools such as voice mail, electronic publishing, image processing, and project management.

Tom Malone and his colleagues at MIT (The Information Lens: An Intelligent System for Information Sharing and Coordination) describe a prototype intelligent system that helps people manage and control, in ways that make sense to them, the electronic messages they receive and send. In addition to electronic mail, bulletin boards, and conferencing, the system framework supports information retrieval, calendar management, and task tracking. The user interface for the system is based on a consistent set of display-oriented editors that expose the underlying knowledge representations in a way that is simple for non-programmers to use and that can be incrementally enhanced by members of a group.

Tora Bikson and her colleagues (Flexible Interactive Technologies for Multi-Person Tasks: Current Problems and Future Prospects) describe an ongoing program of research at the Rand Corporation which seeks to describe empirically how systems actually support work groups and how the work might change as a result. The results indicate that, although the technologies employed by workers in the studies were fairly rudimentary, they did make information tasks within a group more manageable, increase throughput, and permit more broadly-based and flexible work groups. However, the research also suggests that realization of these benefits depends heavily on the resolution of social questions including definition of group norms and values, role structuring, and responsibility for task management—questions that implementors of new technology in organizations are generally not prepared to address.

Calvin Pava (Organizational Architecture for Distributed Computing: The Next Frontier in System Design) of the Harvard Business School examines the organizational implications of the next generation of computer and communications technology. He analyzes two previous architectural domains of system design: subsystem components and user interfaces. He draws a parallel between the development of these domains and the next domain: the development of organizational architectures and computer architectures as reflections of and in support of each other. Pava argues that systems designers of the next generation will need to be educated in organization design and that human resources practitioners and organization designers will play a crucial role in the design of information systems to support their organizations.

Paul Cashman and David Stroll of Digital Equipment Corporation (Developing the Management Systems of the 1990s: The Role of Collaborative Work) envision and put into operational terms a system that will support management coordination. Their thesis is that management systems must deal with a high level of complexity—not by simplifying as in traditional approaches to automation, but by helping to sustainably manage complexity. In their view information technology can help to this end by providing powerful means to structure information and processes of coordination. They describe a prototype Management System they are developing for their own work group at Digital, which helps to structure and retrieve information stored in a variety of types of repositories located on a local or wide area network.

Mark Stefik and John Seely Brown of Xerox Palo Alto Research Center (Toward Portable Ideas) describe in detail current research at PARC studying work group collaboration and supporting technology. The basic theme of their work is the requirement of systems to represent externalized (shared) ideas; their view of this is a shared (electronic) whiteboard. They propose active and sharable workspaces that can support computer-mediated conversations anywhere from offices to coffee lounges to formal meeting rooms. The paper focuses primarily on Colab, an experimental room for meeting augmentation, and the software tools, such as Cognoter, that support collaborative processes in a Colab setting.

The last chapter, by Jon A. Turner and Barry Floyd of New York University (A Method for Evaluating Work Group Productivity Products) proposes a method for evaluating products that can be categorized as "work group productivity support." Their model defines functional performance, administration, and organizational fit as key categories of evaluation applied across the domains of task characteristics, group characteristics, and communication characteristics. The Appendix to their chapter contains examples of use of their evaluation model on four products that were demonstrated at the Symposium.

This set of papers is representative of a significant emerging interdisciplinary field of research as well as significant product development. In some ways it is reminiscent of the emergence of the field of human factors in system design, when cognitive and experimental psychologists began a meaningful dialog with computer scientists and a significant amount of progress in user interface design took place. This body of work recognizes the limitations of the "human-computer" dialog; it acknowledges that knowledge work is not only "personal," but "interpersonal," and thus requires interpersonal computing support. At the same time, rapid technological developments in "connectivity" are taking place, allowing interpersonal computing to be technically feasible. A significant amount of work in both technical design and evaluation of work groups to provide requirements for design remains to be done. We expect to see this new body of interdisciplinary work emerge rapidly and we fully expect to see some exciting results in the near future.

MARGRETHE H. OLSON

1

USER APPROACHES TO COMPUTER-SUPPORTED TEAMS

Robert Johansen
Institute for the Future

Computer-supported teams are small collaborative work groups that use specialized aids. Typically, these are project- oriented teams with important tasks and tight deadlines.

The team members may be present in the same room or they may be attending an electronic meeting at which other participants are not in the same place at the same time. If team members are physically separated, they may decide to use a store-and-forward communications medium that allows them to communicate according to their own schedules. Sometimes, computer-supported teams are permanent groups; more often they are ad hoc task forces with a finite lifetime. The group interaction might be formal or informal, spontaneous or planned, structured or unstructured.

Although computers have been used to support team efforts, the emerging concept of computer-supported teams differs from that of past computer support. Many computer systems, such as timesharing, are already used by more than one person, but such user "groups" are simply aggregations of individuals. Each computer user is seen by the system as a discrete unit; there is little or no direct interaction among the users. Computer-supported teams introduce a new dimension: software designed specifically for groups.

APPROACHES TO COMPUTER-SUPPORTED TEAMS

What can group-oriented software do to support the work of teams? This chapter begins by introducing 17 approaches to team support as it is already beginning to appear. A definite overlap occurs among some of the approaches, but each has its own perspective. These 17 approaches represent a variety of possible steps toward computer-supported teams; the steps get larger (with reference to the present) as the list progresses. Each approach is described, illustrated by a brief scenario, and followed by an assessment of its current status and possible pitfalls.

This is an inductive approach to computer-supported teams: it begins by simply describing what is happening in the user world. After this overview, I categorize current efforts and discuss what is likely to happen next.

1. Face-to-Face Meeting Facilitation Service

Face-to-face meetings are already a way of life in business and there are specialists in facilitating meetings. Typically, facilitators are independent consultants, but large companies sometimes have in-house specialists on call. Today, the normal tools of the facilitator are the flip chart pad and the felt tip pen. What if electronic support for the facilitator were available which, in turn, could support the activities of a work team?

Scenario I "Chauffeur" (Support for face-to-face meetings)

The team members are in a spirited argument as they explore their options for presenting an interim report on their work to date. This is a meeting to plan the presentation they have scheduled with their boss in two weeks, the halfway point in their task force assignment. As the team members talk with each other, a facilitator types quietly at the side, recording summary phrases from each statement that are projected on a screen for the group to see. Periodically, he stops the meeting and asks the group members to check what he has recorded for accuracy; he then tries to organize it into a more coherent whole. Some of the notes created by the facilitator look like electronic versions of what would have been written on flip chart pads. There are also brainstormed lists of ideas and graphic summaries that the facilitator thinks might work for the executive presentation. As the meeting ends, the team agrees on four alternatives for consideration. Draft versions, as well as the complete meeting notes, are printed on a laser printer at the back of the room and photocopied for the team members to take as they leave.

CURRENT STATUS: A small company called Meeting Technologies (Berkeley, California) performs a service similar to the one described in the scenario using three Macintosh computers connected together and some special software for group recording. Several other group facilitation companies are moving in a similar direction. Also, several user organizations have constructed permanent rooms to support such facilitation activities.

POSSIBLE PITFALLS: Facilitators are not well accepted in most companies. In addition, most facilitators are not adept at computer use and the software tools for such facilitation are not yet fully developed. Conference rooms have to be specially equipped to support such activities, or the facilitators will have to carry their equipment, like traveling rock groups.

2. Group Decision Support Systems

Decision support systems (DSS) have gradually emerged and are now used heavily within many companies. Keen and Scott Morton (1978) introduced the concept of DSS as the use of computers to "(1) Assist managers in their decision processes in semistructured tasks; (2) Support rather than replace, managerial judgment; (3) Improve the effectiveness of decision making, rather than its efficiency" (p.1). Why not extend the DSS concept into Group Decision Support Systems (GDSS)?

Scenario 2: "GDSS" (Support for face-to-face meetings)

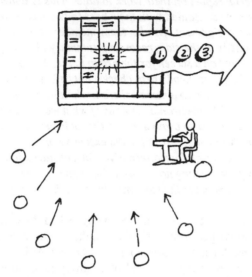

The team has to decide. There are seven different views among the seven team members, but they have to reach one decision. The first thing they agree upon, though not easily, is how to phrase the question, how to decide what they have to decide. Next, the GDSS asks them for anonymous judgments, it asks them about their own uncertainties, and it asks them to self-rate their expertise. After all the team members have entered their judgments, the system does some aggregation of the opinions and feeds back a first set of judgments from the group. The group goes through a series of these "rounds" until a decision is reached. The system certainly does not make the decision, but it provides an effective and efficient group decision-making process.

CURRENT STATUS: Group decision support systems have been in use in limited ways for almost twenty years. Kraemer and King have conducted a recent survey of such systems and conclude that, in spite of years of attempts, "The field of GDSS's is as yet not well developed, even as a concept" (Kraemer & King, 1986; see also Gray, 1986). There are isolated examples, but there is little success to report. However, research activities are increasing and the techniques for decision support are becoming both more powerful and less obtrusive for users.

POSSIBLE PITFALLS: Decision support tools, while plentiful for individuals, often lack the flexibility needed for group applications in business. Furthermore, formal procedures for decision making are often frowned upon by "real business people"; significant changes in perceptions and procedures may need to occur. Conference rooms may need to be adapted to allow for GDSS and this is likely to be expensive. Most companies are used to conference room expenditures that only include such inexpensive items as overhead and slide projectors, or perhaps a speakerphone.

3. Computer-based Extensions of Telephones for Use by Work Groups

The telephone is a "workstation" that is familiar to everyone. If it is possible to build from the telephone, the leap to computer-supported teams does not seem as great for prospective users. There are two basic approaches that are not necessarily exclusive. One builds on the capabilities of the telephone network itself, or private networks; the other builds on the on-premises private branch exchanges that are already common and are becoming powerful.

Scenario 3: "Telephone Extension" (Support for electronic meetings)

The team meeting is booked for 2:00 p.m. and the phones ring right on time. Each team member sits at his or her desk, with a screen display that shows a virtual conference room table indicating who is present and who is talking at the given time. Each of the seven team members is acknowledged on the screen, with their voices coming through the high-quality loudspeaking telephone. When a team

*member has a draft or some data to show, it can also appear on the screens. To
the team members, the system is an extension of their telephones, an extension that
includes what they used to do on the personal computer and what they used to do
through a surly conference call operator. Face- to-face meetings still occur, but
the telephone meetings provide much more regular communications options.*

CURRENT STATUS: Northern Telecom's Meridian already provides
telephone meeting services very much like this scenario, including one called
"Meeting Communication Services." Meridian is a PBX that also acts like a computer. From another perspective, a telephone network-based product that provides
capabilities similar to those in the scenario is the AT&T Alliance bridging service.
This service now provides long distance conference calling for much of the United
States through a digital bridge that also has capabilities for exchanging graphics
among group members. These are both leading edge products, but it is expected
that they will be followed by an increasing number of group-oriented telephone
products and services.

POSSIBLE PITFALLS: PBXs are just developing meeting support
capabilities and these may be tied to expensive purchases of complete new PBX
systems. This linkage to other systems is positive in the sense that group support
capabilities will be positioned as features of the new system, but it may be difficult
for teams to get access to such systems without becoming involved in a larger purchase decision. Network-based services do not have this problem, because they
can be sold as services and prices can be based on use. Telephone-based approaches also need to be connected in some way to the computing equipment already used by teams, and this connection may be difficult.

4. Presentation Support Software

Team members often have to make presentations, either to the team itself or to
others with an interest in what the team is doing. Software can make the process
of preparing presentations much easier, even if the meetings themselves have no
new electronic aids. Instead of relying only on a graphics artist, with frequent long
delays, many presentations can be prepared by the author. Professional graphics
assistance can also be used, but many uses of graphics for teams will not require
such specialized skills.

Scenario 4: "Presentation Prep" (Support for face-to-face meetings)

The team has worked over the ideas for weeks. Now it is time to do a briefing for the boss and the boss's boss. Vugraphs (overheads) are the medium of choice in this company, so the new ideas have to be boiled down into vugraphs. Each team member has played with vugraph content, formats, and styles before the meeting. After going through various drafts, they finally agree upon just the right "look" for their presentation. Then comes the final rush: as usual there are changes up to the last five minutes before the meeting. When it is over, the presentation looks great, except for the laser-printed typo in the lower right-hand corner of the concluding paragraph.

CURRENT STATUS: Presentation software is becoming more common, primarily because of the rise of "desktop publishing." In fact, presentation software is a variant of desktop publishing. One aerospace company has developed software that is geared toward its own internal project briefings, with slide preparation software (for preparation and display on personal computers) and links to conference calling capabilities. Several software companies are introducing extension packages that allow output from existing software, such as a spreadsheet or an idea processor, in a form that can be used directly for presentations.

POSSIBLE PITFALLS: Presentation software may introduce role conflicts within an organization; presenters are not used to creating their own visuals and graphics artists may feel left out. Role changes will need to occur, such as presenters learning enough about style and format to use the software. Graphics artists will need to learn the software and to adapt their skills to those areas where a nonartist with software cannot perform well. In addition, quality control problems can arise, resulting in "laser crud" (Saffo, 1987, p.57).

5. Project Management Software

Work teams have obvious and often pressing needs for task planning and coordination. Specialized software can help them plan what needs to be done, track their progress in reaching goals, and coordinate activities of individual team members. The big issue with project management software is to find a system that all team members will actually use.

Scenario 5: "Team Conscience" (Support between meetings)

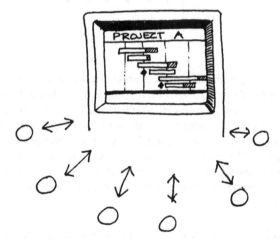

The team has better things to do with its time than keep records. There is a harsh set of deadlines to remember, however. While the team focuses on the content of its work, the system has a basic record of tasks to be conducted, task assignments, subtask breakdowns, and schedules. Each team member reviews his or her progress with the system on a weekly basis; the system is used during team meetings every other week. The software has very little intelligence; it simply organizes what the team has to do and reminds the members when it has to be done. The resulting team discipline and coordination are probably more important than the actual functions performed by the software.

CURRENT STATUS: Project management software is becoming increasingly common on personal computers and increasingly good as well. Some systems even include limited artificial intelligence capabilities that allow for internal judgments about progress or lack of same.

POSSIBLE PITFALLS: Any approach to project management must be used by all key team members in order to be valuable. Project management software must be compatible enough with the styles of team members to allow this participation to occur. This will be tough for software designers, because the needs and styles of work teams will vary greatly.

6. Calendar Management for Groups

This is a straightforward approach: work teams need to coordinate calendars with each other and perhaps others. Unfortunately, implementation of calendar managers is not as straightforward as the concept implies. Many people are reluctant to use computer-based calendars, often with good reason. Yet anyone who has tried to schedule a meeting among several busy people will have thought: There must be better way.

Scenario 6: "Our Black Book" (Support between meetings)

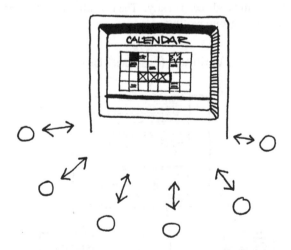

Each team member designates times that are unavailable and available, with a weighing flexibility in the event the system has trouble finding matches or free time. At first, it is hard to get everyone to use the system. Gradually, however, the team agrees that "The Black Book" should be the calendar of last resort and that each team member has to be responsible for keeping his or her own calendar in synch. "If only it would fit in my pocket!" is the recurrent lament.

CURRENT STATUS: Electronic calendars have been accepted very slowly within most user communities, especially by those people who have secretaries or assistants who will schedule meetings for them so they can avoid the hassle. Gradually, however, calendaring systems are coming into the marketplace. On the research side, the logistics of group calendaring are becoming better understood and the implications are promising (Greif & Sarin, 1986).

POSSIBLE PITFALLS: As with project management, group calendaring requires full participation and this will be difficult to achieve on many teams. In addition, many people are very protective of their personal calendars; they are likely to resist the notion of an electronic calendar, especially when it is shared and, to some extent, under the control of others.

7. Group Authoring Software

Group authorship is a common practice already, typically via a series of scrawled comments that are centralized onto one draft before changes are made. Group authorship software would allow team members to make document revisions, with the system remembering who made which changes. Team members could suggest changes without wiping out the original; comparisons among alternative drafts would be allowed easily. The overall goal would be to improve the speed and quality of group writing.

Scenario 7: "Group Writing" (Support between meetings)

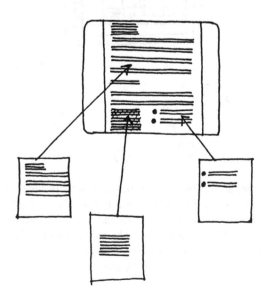

The brief is being filed today in San Francisco because that is where the court is, but the principal attorneys are in New York and Washington. The first draft was done in New York and shipped electronically to Washington and San Francisco. Changes were made in all three cities; the system kept all the versions of the brief, with indications of authorship. The lead attorney made decisions to take this paragraph from Washington, this one from San Francisco, and so on. The brief is being filed on time.

CURRENT STATUS: Group authoring software has been introduced recently by at least five separate companies, all with interesting products (Dalton, 1987). However, these introductions have just occurred and it is too early to see how successful they will be. Some word processing programs are also being expanded to include group writing capabilities.

POSSIBLE PITFALLS: Group writing is a delicate process at best and working together through software could increase the difficulties for some work teams. Even if the software works well, coordination of the various authors will be critical to a successful result. Some teams may give up early because the barriers of behavior change, learning, and coordination are too imposing.

8. Computer-Supported Face-to-Face Meetings

In this case, the team members work directly with computers, rather than through a "chauffeur" as in approach 1. This is a bigger step, of course. There are requirements for more than one workstation in the room, for software that can provide direct group support, and for the users to be skilled enough to use the system directly. It builds, however, on the familiar notion of face-to-face meetings.

Scenario 8: "Beyond the White Board" (Support for face-to-face meetings)

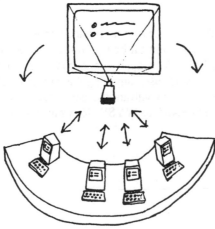

Each team member had been working on a section of the final report. They walk into the specially-equipped room with diskettes in hand (although one person has managed to send his files through the company's local area network from his desktop workstation to the conference room). The half-circle table includes four personal computers connected together and a display screen. Team members work privately during the meeting or display their work for others to see. In the meeting, they work through each section of the final report, doing revisions on the fly. When they leave the room, they leave with a common "group memory" of what has occurred and which steps will occur next.

CURRENT STATUS: The COLAB at Xerox Palo Alto Research Center (PARC) is already beyond this scenario, though only for a single user group. It is based in the Intelligent Systems Laboratory at PARC and is designed for a high-level team of artificial intelligence researchers (Stefik, et al., 1987). (COLAB also includes functions involved in nine other scenarios in this paper.)   Several commercial attempts to develop more limited systems have met with little commercial success. As an example, International Computers Limited (ICL) has a system called "The Pod", which is specially designed for group decision support and is designed in modular units.   Research experience is yielding significant insights that can contribute to future products (Kraemer & King, 1986).

POSSIBLE PITFALLS: The technology for face-to-face meeting support is almost here, but it is difficult and expensive to assemble. Integration of the hardware components is also complex and the software, in most cases, is only available in research laboratory settings. The issue mentioned in Scenario 2 regarding the reluctance of companies to spend money on equipment for conference rooms is also a pitfall.

9. Personal Computer Screen-Sharing Software

If one person can make good use of a personal computer and that person is also involved in team efforts, wouldn't it be useful for that person to "share screens" with other team members? This approach to computer-supported teams builds directly on personal computer use: anything that can be displayed on a personal computer screen could be shared with another personal computer screen. The result has been labeled at Xerox PARC as WYSIWIS (what you see is what I see)(Stefik, 1986a).

Scenario 9: "Screen Sharing" (Support for electronic meetings)

"I think we should move this circle over here and turn the arrow in this direction..." He talks as he moves the circle and redraws the arrow on his PC, but the other team members see it change on their PCs as he does it. They are also connected by conference call to discuss the revisions. They are in a "scratchpad" region of the program right now, but the system keeps track of the drafts and of who creates what. At the end of the meeting, everyone has revised versions on their own PCs.

CURRENT STATUS: There have been various attempts to create personal computer software for screen sharing over the past several years. So far, commercial success has come slowly, but there are definite signs of progress. There are at least two problems: first, it is tricky for users to get the logistics down (to be sure you have the right diskette in the right drive, the right modem settings, and so on). Second, although the idea of screen sharing is immediately attractive to many PC users, it also requires some behavior change. Are there really that many times when you want to share a screen while you talk with someone? Screen-sharing software is a logical "stepping stone" class of products. Screen sharing of specialized teams, such as architects or engineers doing computer-assisted design, seems to be the most likely early application area.

POSSIBLE PITFALLS: Screen sharing is one of those ideas that looks great in principle but that has sticky problems at the implementation stage (Stefik, 1986a). As previously indicated, the logistics of multiple users and multiple screens can be very difficult for both system designers and users.

10. Computer Conferencing Software

Computer conferencing provides group communication through computers. It is the group version of electronic mail. Electronic mail systems are designed for person-to-person communications; filing of messages is done by the individual. Computer conferencing systems are geared toward groups; filing of messages is by group and by topic. Computer conferencing is a logical step toward computer-supported teams: Once communications take place through a computer, other forms of computer aids should be easier to add.

Scenario 10: "Invisible College" (Support for electronic meetings)

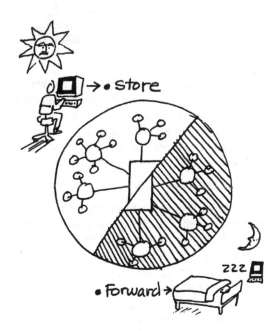

The team is close: They work together each day and often some of them work into the night. The six team members are based in three countries and two states within one of the countries. The "time window" when they are all in their offices is very short. Thus, they usually work in a computer conference. They check the team's conference twice per day, see what has happened since they were there last, make their own comments and leave. Drafts and other working documents, graphics, or models are also exchanged through the conferencing system.

CURRENT STATUS: Computer conferencing has been possible technically since about 1970, but few organizations have taken advantages of its potential. Commercial systems are currently available, but none is doing very well. Several private in-house systems, however, are very successful. It has proved very difficult to get people used to computer conferencing as a general-purpose medium of communication (Hiltz & Turoff, 1978; Johansen, Vallee, & Vian, 1979). The most likely path for the expansion of computer conferencing is through an expansion of existing electronic mail systems to include group communications capabilities.

POSSIBLE PITFALLS: Computer conferencing appears to present more organizational than technical problems. In cases where it has not worked, it has typically been introduced by a forward-thinking management information system (MIS) person who realizes quickly that such capabilities have more to do with a way of organizing work than with a computer system. Computer conferencing can easily create new channels of communication that might be quite different from formal organizational charts. MIS people, of course, typically have little training in organizational change. Thus, computer conferencing is often dropped after a trial, without ever achieving a critical mass of users within a company. Small teams can use computer conferencing without organizational support from their companies, but they must go through independent service providers to do so. Indeed, small teams are the major clients of such services. Most teams simply do not know such options exist, or they find them too expensive or too hard to organize.

11. Text-Filtering Software

Work teams often need large amounts of information that is hard to find. Text filtering allows users to search free-form or semistructured text, with more power achievable through more structure. Typically, users specify search criteria to be used by the filter. Text filtering can also be used to identify people with common interests. In this way, text filtering can be used for computer support of much larger communities, creating a kind of magnet for filtering through text.

Scenario 11: "Needle in a Haystack" (Support for electronic meetings)

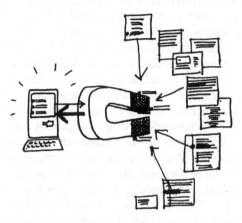

The team uses the filter to search out information and people that will help move its task forward. As is typical with many work teams, the members are working in a field that is still not well understood; they are ahead of the key words in traditional databases. The filter helps them specify just what kinds of information they want. Each morning, the filter prints a personalized "newspaper" for each team member, showing items from the preceding day's news, as well as new findings from the ongoing search for leads. Person-to-person messages are also filtered to insulate the team members from low-priority interruptions.

CURRENT STATUS: Text filtering is being pursued most comprehensively by Tom Malone at MIT(Malone, Grant, Turbak, Brobst, & Cohen, 1987 and Chapter 4, this volume). The original title of this experimental system was "electronic mail filter" and it was intended to help users prioritize their incoming messages. Now his focus has broadened to an "information lens" orientation, whereby the system reaches out to find information that matches the rules created by each user. The Information Lens is a research environment that cuts across the 17 scenarios in this paper; it is directly relevant to twelve of them. Commercial systems for text filtering have not yet begun to appear, but there are definite indications of interest from both users and potential providers.

POSSIBLE PITFALLS: Text filtering is still more vision than reality. While the research is very encouraging, it is still mostly research. The most promising short-run possibilities require much prestructuring of input from users; prestructuring means behavior changes in order to meet the requirements of the system. Possible requirements for such behavior changes are pitfalls to consider if teams pursue text filtering now, instead of waiting for future improvements in capabilities.

12. Computer-Supported Video Teleconferences

Another approach to computer-supported teams is to start with users who are already familiar with video teleconferencing. If they see merit in teleconferencing, it is likely that they will be open-minded about the potential for computer support for the electronic meetings they already hold.

Scenario 12: "Teleconference Assistant" (Support for electronic meetings)

The regular Friday teleconference has just begun and the budget glares back at the team members from the projection screen, with task overruns flashing in red. Each of the two video rooms has four team members present, all of whom are staring at the screens. "What do we do now? We've still got our deadline, but we don't have any money!"

The discussion centers on this question, with periodic recalculations and searches of parallel budgets to come up with additional funds. At the end of the meeting, the numbers are "frozen" for the team members to take along on paper copies. They have to keep working; next week they will decide who will pay for it.

CURRENT STATUS: Computer use within teleconferences has been very low to date. One computer manufacturer, however, uses projections of computer output during audio conferences in a fashion very similar to the scenario above. Also, several video teleconference rooms include personal computers on an experimental basis.

POSSIBLE PITFALLS: Video teleconferencing is still not a usual mode of business behavior. Adding computer support can serve to further increase the sense

of technological discomfort felt by many users. For audio conferencing, the pitfalls can be similar to those noted in Scenario 9 for screen sharing.

13. Conversational Structuring

Communication among team members is a critical aspect of a team's performance, even though thought is rarely given to how to structure this communication most effectively. One approach to computer-supported teams is to develop or
select a structure for team conversations that will be in close keeping with the task
and style of the team participants themselves. Structured conversations might increase both efficiency and effectiveness, if done well.

Scenario 13: "Say what you Mean" (Support between meet-ings)

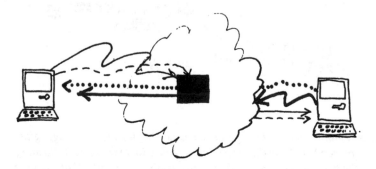

"OK, lets do it."
WHO SHOULD DO WHAT?
"I guess I should get it going."
WHAT WILL YOU DO, BY WHEN?
"I'll do it by Friday."
I'LL PUT IT ON YOUR CALENDAR AND ADVISE THE REST OF THE

TEAM. WHAT, EXACTLY, ARE YOU AGREEING TO DO?

CURRENT STATUS: Conversational structuring is quite a different approach to software. It requires building explicit forms of communication about
what most teams usually do in unstructured ways. The first commercial software
to take a significant step toward conversational structuring is The Coordinator by

Action Technology Corporation of Emeryville, California (Flores & Bell, 1984; Winograd & Flores, 1986).

POSSIBLE PITFALLS: Structuring conversations is risky business. It can be perceived as intrusive or worse. Careful thought must be given to what structures make the most sense for a given team, as well as how to introduce the structures once they have been selected.

14. Group Memory Management

Work teams have an obvious need for a "group memory," particularly if individual members can search the memory in the ways they prefer. The problems arise in structuring data so that it can be retrieved as information by team members. Very flexible indexing structures are needed for this to happen. The term "hypertext" has been used to describe nonlinear indexing structures that allow very flexible storage and retrieval options.

Scenario 14: "Picking Up the Fishnet" (Support between meetings)

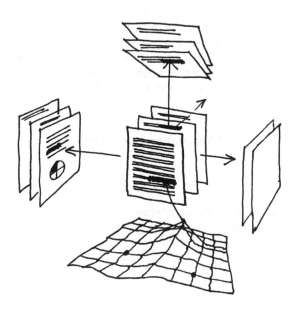

"I remember it was an idea we had a couple of months ago. I think it was Fred, and it had something to do with the notion of 'frequency'." Sara fumes in frustration as she tries to remember the idea.

The Team Memory system contains notes from all the team meetings, with links among many of the words and concepts. Sara follows a weaving and bobbing path through words, data, and people in search of the lost idea. When she finally finds the idea, it isn't nearly as good as she had remembered it. However, the search process triggers a new idea for her, one that is much better than the original one.

CURRENT STATUS: The term "hypertext" was coined by Ted Nelson (Nelson, 1981), but it is only now making its way into regular, though certainly not yet common, usage. At this writing Xerox's NoteCards system is one of the best examples of hypertext (Halasz, Moran, & Trigg, 1987). It is structured around the idea of working on index cards that can be linked and cross- referenced very easily. Also, a hypertext system for the Macintosh, called Hypercards, has been introduced and more such systems are expected to come on the market over the next several years. Hypertext systems have great potential for computer- supported teams. Ideally, hypertext capabilities will become an implicit part of software environments rather than separate software.

POSSIBLE PITFALLS: Hypertext is only now becoming understood and operationalized. Today's systems may be difficult for some teams to access, or difficult for them to use even if they can get to them. Also, this approach to indexing requires the creation of a new infrastructure for at least some aspects of team interaction: It will sometimes take major commitments at the front end in order to create the type of group memory that will prove useful later.

15. Computer-Supported Spontaneous Interaction

It is often said that the most important team meetings happen around coffee pots or in hallways. Can electronic systems be used to encourage and/or support such encounters?

Scenario 15: "Electronic Hallway" (Support between meetings)

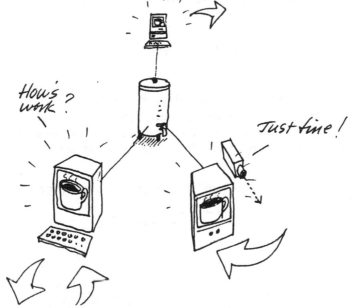

It is almost midnight when Betsy is ready to log off the system. Just then, the system notifies her that Karen has logged on. They type to each other briefly before shifting to an audio link. (Neither of them is interested in a video link at midnight.) A long conversation ensues, the kind that rarely occurs at the office while everyone is rushing about.

CURRENT STATUS: Gordon Thompson (1975) was one of the first to discuss the "electronic hallway" idea and its potential for influencing the formation and operation of groups. The closest manifestation is the System Concepts Laboratory (SCL) at Xerox PARC (Goodman & Abel, 1986). SCL is looking at 5 to 10 years to explore computer-supported group technology and practice. Half the lab is physically located in Palo Alto and half is in Portland. Audio, video, and data links are available between the two groups 24 hours a day. They are emphasizing informal collaborative work for groups of 2 to 10 people. The current system allows "drop-in" encounters over electronic media, much like what currently happens in hallways or around coffee pots. Such communication is very important for teams and it certainly occurs much more frequently than formal meetings in conference rooms.

POSSIBLE PITFALLS: The major hurdles here are logistical: How, specifically, do you go about creating an "electronic hallway"? Today's systems simply are not that portable or that flexible. Thus, the major pitfall can come from expecting too much too soon from this approach. In the long run there is real promise, but it is not clear how long it will take.

16. Comprehensive Work Team Support

Work teams have many support needs and, toward the ambitious end of the spectrum, an integrated computer-based system is certainly attractive. Of course, comprehensive support is difficult to provide, even if the focus is on only one type of team. Still, this is an important direction that is becoming feasible. This is a move toward putting users "inside" their computing environments.

Scenario 16: "It's All Here" (Support between meetings)

The competition is two weeks into a new advertising campaign that is particularly threatening to the brand team. The latest data are now in and it is time to figure out what they mean. Each team member takes a crack at the analysis, sending along draft spreadsheet models and statistical passes through the new data. Finally, they meet around a workstation, with one person doing the updates and final runs.

A summary briefing is then prepared for the brand manager, who receives the briefing and background data on her workstation ten minutes before the meeting at which she is to decide how to respond to the competition.

CURRENT STATUS: The vision of comprehensive team support was first proposed by Douglas Engelbart in the early 1960s (Engelbart, 1963). Engelbart built a prototype system, NLS, that still serves as both a benchmark and a high-water mark. Movement from vision to commercial reality has been slow, however. A commercial version of NLS is available from McDonnell Douglas as AUGMENT. The most significant step to date is focusing on brand teams in packaged goods industries, much like the scenario above. Metaphor Computer Systems, of Mountain View, California, has an integrated system targeted specifically toward these types of high-performance teams. At this point, it seems reasonable to conclude that comprehensive team support can be provided best if it is geared toward specific types of teams, as with Metaphor.

POSSIBLE PITFALLS: With today's products, users are likely to find that the specific functionality they achieve within an integrated system is not as powerful as that same functionality in a stand-alone system. This presents an unfortunate trade-off between the values of integration and power within specific functional areas. This trade-off is not inevitable; it is certainly possible that integrated solutions can be more powerful than their stand-alone counterparts. In addition to this trade- off, integrated systems are also likely to be expensive and are probably not compatible with the mainstream software marketplace.

17. Nonhuman "Participants" in Team Meetings

At some point, computer programs should be able to function, in some sense, as team "members." This is the most ambitious approach to computer-supported groups in the list of 17 and it relies heavily on developments in artificial intelligence.

Scenario 17: "Nonhuman Participants" (Support for electronic meetings)

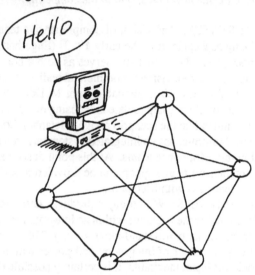

The team meeting for new brokers is just convening. Each trainee has spent the better part of the preceding day working with the Coach, an expert system that has specialized expertise about investment options that the new brokers will be selling in another three weeks.

There are many opinions; even the Coach is only expressing an opinion. The new team discusses the options, consulting again with the Coach at several points during the meeting. The Coach has specialized knowledge that nobody on the team has, but it does not have definitive answers. It is a collaborative process, with all the team members, including the Coach, contributing.

CURRENT STATUS: There are no real examples of a computer program functioning as a team member, although there are several examples in user organizations (all in confidential settings) where similar ideas are being pursued. More detailed scenarios exist that introduce the concept and explore some of its implications (Johansen, 1984). There is also growing interest among artificial intelligence researchers in the role of expert systems as a "knowledge medium," whereby people communicate through an expert system, rather than simply extracting information from it as an autonomous system (Stefik, 1986b).

POSSIBLE PITFALLS: Designing systems that are thought of as people or team participants may be quite misleading and perhaps counterproductive. Today's state of the art means that systems are a long way from personhood; this may always be so, or at least it is likely to be so for a long time to come. Care must be taken that the "nonhuman participant" is not oversold or misunderstood by human team members. In short, expectations must be managed within the realm of realism.

UNDERSTANDING WHAT IS GOING ON

What patterns can be seen across these 17 approaches to computer-supported teams? These approaches obviously cover a wide range of user activities. Although it is too early to conduct a detailed analysis of current products, it is useful to try out various categorizations of the current activities. Table 1 presents all 17 Jpproaches arranged in approximate order of increasing difficulty.

Looking across the user approaches summarized in Table 1, it is possible to do a number of different groupings. The most useful method initially was to categorize the approaches according to one of the fundamentals of any work team, the meeting. A broad definition of "meeting" is meant here, including any form of group interaction. The 17 approaches to computer-supported teams can be classified in the following fashion: support for face-to-face meetings, support for electronic meetings, and support between meetings. Table 2 presents this grouping.

TABLE 1
17 User Approaches to Computer-Supported Teams

1. Face-to face meeting facilitation services..."Chauffeur"
2. Group decision support systems..."GDSS"
3. Computer-based extensions of telephony for use by work groups..."Telephone Extension"
4. Presentation support software..."Presentation Prep"
5. Project management software..."Team Conscience"
6. Calendar management for groups..."Our Black Book"
7. Group authoring software..."Group Writing"
8. Computer-supported face-to-face meetings..."Beyond the White Board"
9. Personal computer screen-sharing software..."Screen Sharing"
10. Computer conferencing software..."Invisible College"
11. Text-filtering software..."Needle in a Haystack"
12. Computer-supported video teleconferences..."Teleconference Assistant"
13. Conversational structuring..."Say What You Mean"
14. Group memory management..."Picking up the Fishnet"
15. Computer-supported spontaneous interaction..."Electronic Hallway"
16. Comprehensive work team support..."It's all here"
17. Nonhuman participants in team meetings..."Nonhuman Participants"

TABLE 2

Categorizing the 17 Approaches to Computer-Supported Teams

Support for face-to-face meetings

Scenario 1:"Chauffeur" (facilitation services)
Scenario 2: "GDSS" (group decision support)
Scenario 4: Presentation Prep" (presentation support software)
Scenario 8: "Beyond the White Board" (commuter-supported meetings)

Support for electronic meetings

Scenario 3: "Telephone Extension" (extensions of telephony)
Scenario 9: "Screen Sharing" (personal computer software)
Scenario 10: "Invisible College" (computer conferencing)
Scenario 11:"Needle in a Haystack" (text filtering)
Scenario 12: "Teleconference Assistant" (for video teleconferencing)
Scenario 17: "Nonhuman Participants" (on-line resources)

Support between meetings

Scenario 5: "Team Conscience" (project management software)
Scenario 6: "Our Black Book" (calendar management software)
Scenario 7: "Group Writing" (software)
Scenario 13: "Say What You Mean" (conversational structuring)
Scenario 14: "Picking Up the Fishnet" (text filtering)
Scenario 15: "Electronic Hallway" (spontaneous interaction)
Scenario 16: "It's All Here" (comprehensive support systems)

Of course, some of the scenarios could fall under more than one heading in Table 2. Scenario 13 (conversational structuring), for example, could fit under any of the three headings: support for face-to-face meetings, electronic meetings, or between meetings. It was put in the last category because that is where it focuses from a user point of view. That is, conversational restructuring can help a team keep itself organized and on track with a structure that moves beyond specific team meetings.

Table 2 suggests that most of today's approaches to computer-supported groups focus on electronic meeting support and support between meetings. These approaches also tend toward the middle and upper end of the difficulty spectrum, as indicated by scenarios 1-17. Two of the options for face-to-face support are at the "easy" end of the spectrum (face-to-face facilitation and group decision support), as are one for electronic meeting support (building on the telephone) and three for support between meetings (calendar and project management and group writing).

Classifying approaches to computer-supported teams according to support for team meetings is, of course, only one possibility. Others to consider (all of which are beyond the scope of this chapter) are the type of support provided, synchroneity or asynchroneity of the group communication that occurs, the size of the group to be supported (in this chapter, only small work teams have been considered), or the type of group to be supported. The purpose of this chapter is simply to introduce the concept of computer-supported teams and to begin a consideration of how the concept might develop in the future. A recent article has introduced useful categorization frameworks from a group decision support systems point of view (DeSanctis & Gallupe, 1987).

FORCES FOR COMPUTER-SUPPORTED TEAMS

The 17 approaches to computer-supported teams indicate an energy around this general concept, even though the concept is obviously not focused. Several forces in the business world are helping to generate this energy for computer-supported teams.

First, and probably most important, is the general trend toward business teams. Teams have become the order of the day for many large companies. Cross-organizational groups are most common: project teams or task forces that have important mandates and tight time deadlines. These groups are searching for tools that will help them get their jobs done. Computer support often seems to provide an answer, or at least part of an answer. Most business teams also have access to funding for such computer tools should they prove useful. In addition, a purchase by a business team will be evaluated by different criteria than a purchase from a data processing, MIS, or telecommunications manager. Although operational people typically worry about saving money, business teams often focus on ways to make money. Consequently, they will often be more willing to try something new and to take promising risks.

As a side benefit, the performance of business teams is more often tractable than the performance of large organizations. If a team has a clear task and a timetable, its productivity should be measurable (or at least more measurable than many parts of a business). This measurability of effects should make business teams even more attractive to executives. Also, if business teams use computer support effectively, the success stories should be convincing indeed.

The only downside from the business teams phenomenon comes from possible excesses. It is easy to imagine situations in which so many team meetings take

place that little real business gets done. Business teams have to be used selective-
ly in high- leverage areas in order to be effective.

Second, but still very important, is the acceptance by most businesses that computers can be used to gain competitive advantage. This notion has been promoted
by the good business schools for over five years, but now it is finally having an effect on the masses of U.S. business people. Because of this realization, there should
be more receptivity toward the idea of computer-supported teams.

Third, the penetration of personal computers has now grown to the point where
interconnection of team members at their desks is usually practical. The evolution
of business-oriented software supports this trend. Those PCs sitting on desks, even
if they are not being used, represent potential building blocks for equipping computer-supported teams. Many business people also seem to have a sense that their
companies might have moved too quickly in the race from the mainframe to the
isolated desktop. Users often realize that they want their PCs to be linked at least
to the PCs of their closest coworkers. Fueling this trend is the increasing popularity
of local area networks.

Finally, the personal computer industry is expressing strong interest in group-
oriented software. Although much of this interest is within research laboratories
of large computer manufacturers rather than product planners, software developers
are also expressing strong interest. One driver for them, of course, is the search
for the "next 1-2-3," the next software bonanza that will spur sales and use of personal computers. The climate in the personal computer industry is becoming more
favorable for the growth of group-oriented software.

BARRIERS TO COMPUTER-SUPPORTED TEAMS

Until this point, I have emphasized the promises of computer-supported teams.
But there are barriers to consider also. The most basic problem is typical of emerging technologies: "It" (computer-supported teams) is very hard to name and
describe quickly to newcomers. Notice the format of this chapter. I went through
many drafts trying to come up with a crisp initial definition. Finally, I resorted to
leading with examples of computer-supported groups rather than definitions.

As a researcher who has worked often with emerging technologies, I have
come to realize that the problem of names is recurring. By definition, an emerging technology is hard to name. If it has a good name, it will not be emerging, it
will have emerged. When an emerging technology has a "grabber" name, its emergence will come much more quickly.

This "unnamability" is tough on prospective users of computer-supported
teams. New ideas are a lot easier to grasp if they have easy "handles" to describe
them. In the case of computer-supported teams, the handles are pretty awkward.

Here are 14 competing terms that mean about the same thing as, or at least over-
lap considerably with, "computer-supported teams":
  -- "Technological support for work group collaboration"
  -- "Computer-supported cooperative work"
  -- "Collaborative systems"
  -- "Workgroup computing"
  -- "Group decision support systems (GDSS)"
  -- "Interpersonal computing"
  -- "Departmental computing"
  -- "Augmented knowledge workshops"
  -- "CAC (Computer-assisted communications)"
  -- "Group Process Support System"
  -- "Teamware"
  -- "Decision Conferences"
  -- "Coordination Technology"
  -- "Flexible interactive technologies for multi-person tasks"

These names all have their virtues, but they are not names that will stir the
hearts of business people.

Just after the Symposium for which this chapter was prepared, Fortune
published one of the first popular business press articles on computer-supported
teams, including what is perhaps the generic name for the field that will be most
attractive to business users: "groupware" (Richman, 1987).

Unless prospective users can agree on a good name, computer- supported
teams should not be billed as something new. They should be positioned as a way
to get done whatever a particular team has to get done. This approach may be best
in the long run as well, because it is not clear that the range of approaches described
here as computer-supported teams will remain part of an integrated field of activity.
As usual with emerging technologies, the early stages involve much uncertainty.

The second barrier to computer-supported groups is even more down to earth:
Group-oriented software is not easy to develop. Most of the problems are nitty-
gritty, rather than state-of-the- art technological, but they are problems nonetheless
(Greif & Sarin, 1986). Thus, the growth of group-oriented software will be
tempered by the difficulties of creating this software. The transition from individual to group software is major and software designers will have much to learn
in the new world of group support.

Third, there are few success stories to date regarding computer-supported
teams. The 17 approaches provide a good taste for the present range of experience,
but user experience is limited. Most of the approaches are not well tested by users.
Furthermore, there are incentives not to exchange success stories with other users
when they do occur. Because teams are often working on important tasks, sen-

sitivities crop up regarding competitive advantage. Many of the user examples I gathered in researching this chapter were described to me under nondisclosure.

Finally, product groups in the major computer manufacturers (as compared to research laboratories within those same companies) are not yet interested in group-oriented products. This is understandable. For most vendors, particularly in these days of short-term financial pressures, there is little interest in products that require customers to make conceptual changes. Big vendors do not make money selling conceptual change, or so the argument goes. At this point, research groups in major computer vendors are pursuing group-oriented products, but the transfer has not been made to the product planners. This transfer will occur, but how long will it take?

A FORECAST

In conclusion, I venture to forecast how computer- supported teams are likely to develop over the next few years. My estimate is that group-oriented software and systems will be rare for the next three years, followed by a period of rapid growth in the three to five year timeframe. The barriers are too imposing to expect rapid growth in the near term, even though the forces for computer-supported teams will win out in the long run.

In the meantime, I expect that innovative user organizations will see the three to five year lag as a "window of opportunity" for them to gain competitive advantages through the use of computer-supported teams. In particular, I expect that user organizations with the following characteristics will be major users of computer-supported teams in the near future:
-- Companies with many decentralized project teams
-- Companies with a high penetration of PCs and local area networks
-- Companies with successful teleconferencing systems
-- Companies with flexible organizational structures
-- Companies with a track record of early adoption of information system
   innovations.

In this three to five year timeframe, I also expect that small software companies will be active. These will be high-risk ventures, however, because they will be leading the behavior change efforts implied by group-oriented software. Small service providers should be quite successful, because it will be feasible for them to package some of the benefits of computer-supported teams and to sell them as a service to users who do not want to undergo major development efforts themselves. The major computer manufacturers are not likely to take the lead; they will wait for the acceptance of group-oriented software. The major software providers have real opportunities in the short run, but they are most likely to let the small companies do the software research and development and test-marketing for them. The successful small providers of group-oriented software will then become candidates for acquisition.

In short, groupware, or whatever it is called, will happen in a big way. The only question is when. For the next three years, I predict there will be big wins by only a few players: innovative users, service providers, and small software companies. In the three to five year timeframe, computer support for work teams will become much more accessible and much more heavily used.

2

ASSUMPTIONS UNDERLYING SYSTEMS THAT SUPPORT WORK GROUP COLLABORATION

Vasant Dhar
Margrethe H. Olson
Information Systems Area
Graduate School of Business Administration
New York University

INTRODUCTION

Computer-based systems that support managerial work have implicit to them a set of assumptions about those aspects of work being supported that are important, and those that are incidental. These assumptions shape the ontological primitives of the system, that is, the models that are visible to and/or manipulated by the user. A significant problem with many systems is that their underlying set of ontological primitives do not match those required by the user. This can have the effect of either over burdening the user with complex task details or forcing the human/computer interaction into a mode that the user finds awkward. In designing systems it is therefore important that designers make appropriate assumptions about what users (i.e., managers and professionals) do, what aspects of work they find problematic, and the role of computer-based support.

Currently, a new class of systems for supporting *work group collaboration* is emerging. Depending on their perspectives, designers of such systems seem to make differing assumptions about what is important in collaborative work. In this chapter we describe two broad classes of collaborative work support systems discussed in the literature and the ontological primitives they incorporate. We demonstrate, via a case scenario, some aspects of collaborative work that such systems do not support, and describe a model that might support these aspects of work via a broader and more appropriate set of ontological primitives.

WHAT IS COLLABORATIVE WORK?

We use the term *collaboration* to refer to a goal-oriented process involving contract definition and execution among two or more individuals. In general, collaborative work requires *communication* and *problem solving*.

Communication refers to the exchange of information for purposes such as notification and clarification. Problem solving refers to the *processing of information* for purposes such as planning, monitoring, negotiating, and decision making (Fikes, 1982).

It is our view that most current systems tend to over emphasize the communication component view to the exclusion of problem solving, or vice versa.

For purposes of analysis, we consider collaborative work to be *project oriented*. A project involves activities that are to be performed by two or more people. The order in which these activities are to be attended to is referred to as a *plan*. The process of synthesizing a plan, particularly when there is discretion in when and how activities are to be performed, involves *negotiation* among the parties. In addition, projects are generally monitored and decisions are made on how to modify or achieve goals as the project evolves. In summary, planning, plan monitoring, negotiating, and decision making can be considered as the basic components of collaborative work.

For a project to be successful, there is likely to be some extent of contractual agreement, however loose, that defines expectations. It would seem then, that some of the same types of difficulties that arise in defining and executing contracts would apply to collaborative work as well. In the economics literature, several factors have been identified as influencing contracts; two of these, *uncertainty* and *complexity* (Williamson, 1975), are useful in understanding "collaborative" contracts as well. In addition, we define a third factor, *ambiguity*, that is an important factor in work group collaboration.

Complexity

In problems that require exploring a space of alternatives, complexity is associated with the size of the search space. A commonly cited example of a problem involving a high degree of complexity is chess where a game can involve a space of 10^{120} possibilities—much too large to contemplate exploring exhaustively, even with the fastest of processors.

In collaborative work, we identify two forms of complexity. The first form arises in projects where the activities to be performed and the resource requirements associated with these activities *are known*. What is of interest to a project manager are the consequences of perturbations in sequencing or resource changes. For example, in a large project modeled as an activity network, complexity involved in assessing the repercussions of changes in time and resource estimates can be significant (although clearly not intractable).

The second form of complexity is important from an individual's perspective; this is the complexity involved in attending to several projects simultaneously. Frequently, projects compete in terms of priorities, making it difficult for the individual to plan commitments appropriately. In particular, if the individual is bombarded with many requests simultaneously and the requests are difficult to index to specific projects, they can be forgotten altogether (Neches, 1986). Finally, individuals may also find it difficult to assess the impacts of changes in one project on other projects with which they might be involved.

Uncertainty

Uncertainty refers to lack of knowledge about what states of nature will prevail. When the possible outcomes of an event or action are too numerous to measure or when the risks associated with events or actions are too hard to define, it becomes impossible for parties to fashion a firm contract (Meade, 1971). Also, when the environmental uncertainties become too numerous to consider systematically, the information processing capabilities of the parties can be exceeded; that is, complexity becomes significant (Williamson, 1975).

From the individual's standpoint, uncertainties are associated with time estimates of projects' activities that the individual is involved in. Also, the uncertainty of others' multiple commitments make it difficult to plan and monitor actions, and to negotiate firm contract requirements.

Ambiguity

In the preceding discussion, we assumed that the project activities, regardless of the uncertainty associated with them, were well defined. This may not be the case, particularly in the early stages of a collaboration. Specifically, even though collaborative work may be characterized in terms of contracts and commitments about tasks (activities), determining the meaning of such tasks and hence the commitments they entail may be fuzzy. Suchman and Trigg (1986) and Kraut, Galeger, and Egido (1986) have observed that collaborative research relationships can go through several phases such as apprenticeship, working relationship, and reward division. They note that in the early phases, expectations may be deliberately fuzzy to allow the parties to adjust to the relationship. Thus, even though two parties might agree that some activity is to be performed, there might be only sketchy and partially overlapping understandings of what the activity entails, and what the criteria for evaluating performance might be. In this type of collaboration, ambiguity is resolved during the course of the project. The bulk of this resolution of ambiguity might take place during task definition. Performance criteria and expectations can also change during the project in situations where one party convinces the other that some variation of the originally agreed upon task is also acceptable.

It should also be noted that the degree of cooperation among individuals in a collaborative work group can also vary. We speculate that the degree of cooperation within the group can affect the rate at which ambiguity is resolved. For example, if the degree of cooperation within the group is low (e.g., because of competing goals), there might be persistent attempts at convincing other parties to modify the task or expectations. Ambiguity regarding definition of contracts may therefore persist.

In the following section, we describe a project involving the development of a large computer-based information system in an insurance company that one of the authors was involved in. This case illustrates how the three factors described above affect contract definition and execution in collaborative work, and what aspects of such work computer-based tools do and do not support.

A Case Scenario

The project described here is a fairly typical type of project involving work group collaboration. The organization in which the system was developed is a large insurance company specializing in "group" benefits. The company is divided into two major functional divisions: insurance and pensions.

The system development effort involved both divisions. Specifically, the organization wanted to develop a line of "full-service" products for smaller corporate clients where the client could access information on pensions and insurance directly from the insurance company's computers and generate reports whenever desired. In effect, the insurance company was developing software packages to enable a client to perform information and transaction processing for both (pensions and insurance) functions. Because of the scope of the product, its development required a considerable amount of expertise from both divisions.

The project team consisted of six members, three from each division. All members were also working on other projects at the same time. There was a designated project leader from the insurance division reporting to insurance division management with no reporting responsibility or authority on the pensions side and a similar designated project leader on the pensions side of the project.

As originally conceived, the project was to go through the following phases:

Information requirements definition. There were two types of interdependent requirements:

General product requirements
Individual application requirements

Specifically, the client would purchase a general "turn-key" system of hardware and software for data access, report generation, and answers to specific queries. The client would also choose from a number of available (pensions and insurance) applications, depending on what services the client wished to use.

Software package acquisition/System design. The company decided to purchase a package from a software vendor that provided much of the functionality required for the general product requirements. The package was for running programs residing on both a personal computer and a mainframe (which would be the company's on-site computer), accessing a mainframe database, and presenting a color graphics interface on the personal computer. The location of processing (mainframe or personal computer) was to be invisible to the user (the client). The company expected that the software package would need to be modified slightly to meet their needs and that the package vendor would take care of the modifications.

Application development: Pensions and Insurance. Actual development of the applications (pensions and insurance) was to be carried out separately by members of the two subgroups. It was assumed that the product requirements, which provided the constraints on application requirements, would be well defined, so that communication between the two subgroups during this phase would be minimal.

Integrated user interface design. The interface, the part of the system which the client actually saw, had to be consistent across applications. Although the design of the user interface was part of the general product requirements, it had to be designed and tested with the applications. Issues such as the structure and wording of commands, use of color, and spacing had to be resolved.

Implementation and maintenance. This phase entailed bringing the applications together on the "turn-key" system, designing a "master menu" for accessing the product, and performing system tests. It also entailed delivering the product to its (beta site) clients and monitoring its usage for necessary modifications.

Figure 2.1 shows a schematic of activities involved in the various sub-projects. Specifically, the five phases outlined here have been decomposed hierarchically into lower level activities and precedence relationships between them have been specified. As a caveat, we should point out that not all the activities and their

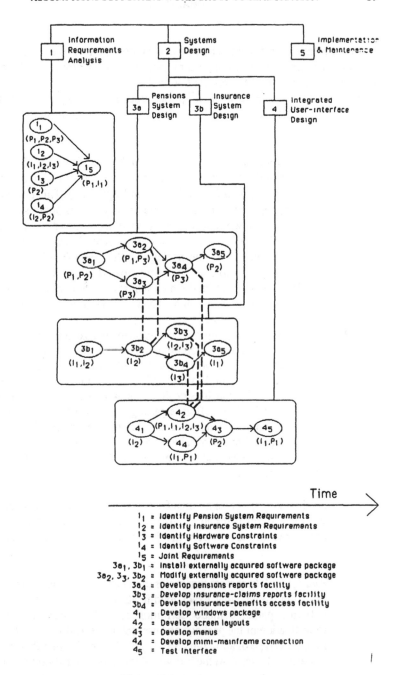

Figure 2.1

precedence relationships were definable at the outset of the project. In fact, as we pointed out earlier, one of the important aspects of work group collaboration in general is the definition of the relevant activities.

As it turned out, one of the assumptions crucial to many parts of the project did not hold up. Specifically, the following events occured during the course of the project:

- The software package purchased from the outside vendor did not work as anticipated (affecting the tasks involving modification of the package—tasks 3-(2) 3a-(3), and 3b-(2) in Figure 1.1). This event had the following repercussions: there were considerable problems identifying what additional work needed to be done in order to make the package operational. Furthermore, there was considerable disagreement as to whether the vendor of the package or the client (the insurance company) was responsible for the unanticipated work. Within the group, it had not been determined which subgroup should have responsibility for the modifications, because it had been assumed originally that the vendor was responsible.

- Individuals in the group were called on to change their priorities related to other commitments, potentially affecting their plans and ability to meet deadlines. However, it was difficult to assess the details.

- A number of tasks (4(-1), 4(-2) and 4(-3) in Figure 1) required acquisition of new skills (such as designing new types of graphics interfaces using the unfamiliar software package) that to begin with, were factored into approximately the original time schedule. With the new problems posed by the purchased software package, some members of the team were unsure about how to modify their commitments to this and other projects.

Although this description is a considerably simplified version of the actual scenario, the events described can be used to illustrate how the three factors described in the previous section were involved in contract definition and execution:

- *Ambiguity.* There were two major sources of ambiguity in the case. Not knowing about what exact modifications would be required to the software package made definition of tasks, particularly those of modification of the package and user interface, difficult. At the project level, this type of ambiguity also left unclear which members were best suited for the task. At the individual level, it was difficult to plan commitments to this and other projects. When expectations about the software package were not realized, the ambiguity associated with the software modification tasks (and those following them) increased.

- *Uncertainty:* The external software package required that the originally uncertain time estimates associated with the applications and user-interface design tasks be revised. However, generating meaningful estimates required that the ambiguity associated with the newly recognized tasks (modify the software package) be resolved first.
- *Complexity:* Complexity is only apparent once the activities, responsibilities, and time estimates have been defined. Particularly at the individual level, this would make it possible to assess the individual's statuses (commitments, slack) on other projects.

EXISTING SUPPORT TOOLS

So far, we have said nothing about existing computer-based systems for supporting collaborative work. In this section, we describe two types of computer-based tools designed to support collaborative work. Their potential role and limitations for the type of problem described in the case are also described at the end of this section.

A Project Management System

Most project management systems are based on the Critical Path Method (CPM) model. The primitives provided by such systems are activities, resources that can be used to carry out the activities, and precedence relationships among the activities.

Such systems are driven by a set of assumptions that we refer to as the *problem-solving view*. The assumptions are:

- All activities are known and relatively well-defined at the beginning of the project, and time estimates are available. There are well defined precedence relationships among the activities.
- The primary factor being addressed is *complexity*. Negotiation, if any, is "outside" the model, and pertains to time estimates of activities, not the deliverables of the activities. Each project that an individual might be involved in is modeled separately.
- A secondary factor being addressed is *uncertainty*. The model can be run with varying time and resource constraints and objectives as new information becomes available.

In the case scenario, the project management system could be used in the following way. At the beginning of the project, all relevant activities, as well as time and resource estimates and constraints associated with them, would have to be defined. The model would be run with various estimates and objective functions. Assignments of individuals to activities would be made. Based on this assignment, an initial resource allocation plan would be derived. As group members review their assignments and make alternative suggestions ("John is better at graphics than I am"; " I don't want to deal with the user - we don't get along"), the model could be used to assess the effect of reassignments on the overall plan. In this way, the system could handle the complexity of assessing changes at the project level once the constraints are specified.

As system development continues, uncertainty in task definition could force changes to the plan. For instance, when the purchased package turns out not to perform as expected, a partial restructuring of activities would be required. After redefining the project, the model could be rerun with various scenarios as done initially.

Periodically, the project management system could also be queried by management for project status. In particular, a project leader might be interested in knowing about changes in the critical path and potential resource shortages.

An Electronic Mail System

Unlike project management systems, there is no standard model underlying electronic mail systems. However, almost all such systems have standard features for sending, replying, and forwarding messages. More sophisticated systems give users added customization capabilities for filtering and prioritizing mail (Malone, Grant, Turbak, Brobst, & Cohen, 1987). More sophisticated systems usually have a relatively sophisticated filing structure; i.e., an indexed external memory; individuals can also cross-reference messages using user-defined keywords.

The electronic mail system is driven by a set of assumptions that we refer to as the *communications view*.

- Not all individuals have all the relevant information about the group activities, and a primary communication task is to force the parties to resolve *ambiguity* and to some extent uncertainty through the sharing of appropriate information. This is accomplished through making explicit requests (i.e., task assignment) and commitments (i.e., acceptance of task assignment). For the tool to be effective, it is assumed that via open-ended communication, task ambiguity should have been reduced by the time commitments are made; both parties have a "shared meaning" of the task that is being committed to.

- The individual can somehow resolve *complexity*, particularly if involved in several projects, to the point of making commitments to requests.
- Electronic mail provides added value over verbal (i.e., face-to-face or telephone) requests and commitments because it provides a written record after-the-fact. It provides added value over written memos because it is quicker and more action-impelling.

In the case scenario, the electronic mail system could be used during the life of the project to clarify responsibilities and the status of various activities in the project. Its primary role would be to keep the members of the group informed about project status, and to reduce task ambiguity and uncertainty via communication.

Critique and Recommendations

Abstractly, the collaborative process is a network of conversations involving message passing among the individuals in a group. As Winograd (1986) points out, each message is essentially a *request* directed for some *purpose* (or action). The fundamental limitation of many communications-oriented systems is that they fail to provide any decision-making support to the individual, that is, a basis for responding to the request given existing commitments. In order for this type of support to be available, a computer-based system must be equipped with a problem-solving model that is capable of assessing the ramifications, for both a project and an individual, of the request for action. Ironically, many project management tools do provide a model; however, such existing tools provide little support at the individual level.

In the remainder of this section, we critique the two types of tools, and propose a scheme whereby both the communication and problem-solving aspects of work group collaboration might be supported. [1]

---

1   The description which follows takes some of its ideas from products which are in use today and are familiar to the authors, (i.e., the Project Workbench of Applied Business Technology Inc. and the Coordinator of the Action Technologies Corporation). We consider these as prototypical project management and communication tools respectively.

Basic communication tools such as electronic mail provide little for dealing with the complexity of the interactions an individual might be involved in. When the network of conversations one is involved in becomes complex, this lack of support works against the usefulness provided by the system. Enhanced communication tools such as the COORDINATOR (Winograd, 1986) and CHAOS (de Cindio, DeMichelis, Vasallo, & Zanaboni, 1986) take the view that the language used for contract formulation and negotiation cannot be analyzed in terms of its semantic content, but can be characterized in terms of speech acts, namely, directives, commissives, and declaratives. In effect, they sidestep the issue of interpreting the "meanings" of requests and commitments, and instead deal with the structure of the conversation, which in collaborative work can be expressed in terms of these acts. The acts serve to impose a structure on the conversation, making explicit to the user the history of the conversation, the categories of messages involved, and standard responses to various categories of messages. For example, an individual might react to a *request* (a message type) by rejecting it, countering it, promising to do it, or withdrawing a previous commitment (standard responses to the request).

The major limitation of communication support tools is that they do not provide support for what individuals *should* do, or the consequences of their actions. For example, in the case described in the last section, when the package purchased from the vendor did not work as planned and required modification, a redefinition of projects and a reallocation of resources to activities was required. The type of support needed in such situations must be able to help expose the more plausible courses of action within the existing set of commitments. This enhanced functionality requires a problem-solving component that is cognizant of the individual's commitments as well as others', and that can use this knowledge to inform the individual of changes that are occurring at the project level and the potential impacts of these changes on their plans. Unaided, people are often swamped by the complexity and uncertainty involved in existing commitments to the point where determining appropriate responses to new requests becomes difficult.

On the other hand, project management tools are capable of assessing the consequences of changes in resources and time estimates. As we pointed out, however, such tools have not been utilized at the individual level. For example, when changes occur, they do not enable a project manager to answer questions such as: Who has both skills and time to modify the package? Is it reasonable for the individual to accept a new or modified reassignment? What reallocation is likely to be least detrimental to overall scheduling?

In summary, by limiting their primitives to project activities, project management systems do not capture other important types of knowledge in projects. They do not model the network of commitments that an individual is involved in. Our view is that the project management system framework can be applied to the end level to model commitments and constraints, and that the analytical capabilities provided by using their models would provide useful input into communications oriented systems.

TOWARD AN INTEGRATED TOOL TO SUPPORT COLLABORATION

We view an enhanced support tool as one that can base decisions about communication primitives (e.g., speech acts) on a problem-solving component capable of assessing consequences of actions. Figure 2.2 shows a schematic illustrating responsibilities of individuals to projects. Specifically, individuals are explicitly linked to activities and on-line records of their plans. For clarity we have shown only a few links.

An integrated tool must have the following functionality:

- enable an individual involved in multiple projects to *plan* actions, that is, set up a temporal network of actions;
- monitor projects in order to keep individuals informed about the status of activities and changes in existing estimates, and to use this knowledge to indicate potential plan modifications at the individual level;
- monitor individual actions, assess when the likelihood of delivering on promises might begin to decline and the impacts of changes in an individual's plan on projects and other individuals' plans.

Enhanced with this functionality, the project management system could become useful as an individual and project planning tool. A person can plan, or at least anticipate, critical deadlines and priority conflicts. The schedule of tasks can be represented weekly, daily, or even hourly (in the case of multiple tasks of very short duration and very fast turnaround). The explicitness of the record itself helps the individual to plan, just as making a list helps one see clearly what needs to be accomplished. Thus the system deals with complexity of both the project and individual activities and priorities.

As we described earlier, ambiguity arises for two reasons: the task itself is not well-defined, or competing interests preclude task expectations from being made explicit. To some extent, dealing with complexity should reduce ambiguity of the second type. When the individual's commitments (plans) are generated and represented explicitly, they provide the individual with more information that can be used to respond knowledgeably to a request. The power of the electronic mail component is in making requests and commitments explicit, thus reducing the potential for ambiguity in task assignment.

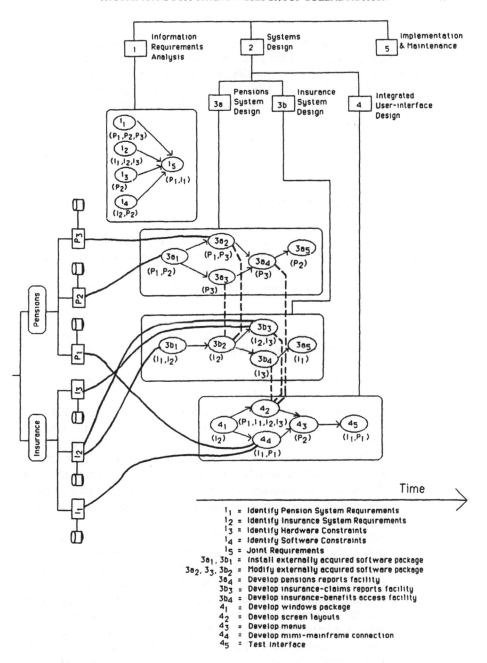

Figure 2.2

In terms of a broad design, the integrated model we have outlined must link individuals and projects, with the potential to set up individuals' plans based on project requirements or to set up/modify projects based on individuals' commitments. Depending on how such a system is to be used, some of the links can be disabled. In this way, it can be used to support models of "collaboration" ranging from strictly hierarchical (directives flow down, information flows up) to completely participatory (both directives and information flow in all directions). The amount of information about other projects or plans available to any one individual or project may also be varied.

THE CASE SCENARIO REVISITED

How might the system be useful for the type of case we have described? Initially, there may be an original set of activities that is generated by the co-project managers about assignments of individuals to sub-projects. A system that integrates the communication and project management functionalities could be used to do the following:

- Establish communication channels among relevant individuals. In the case, communication was needed between activities 3a-(2), 3a-(3) and 3b-(2), that is, individuals P-(1), P-(3), I-(2). Communication was also needed between activities 3b-(3), 3b-(4) and 4-(2), that is, individuals I-(1), I-(2), I-(3) and P-(1).
- Enable all individuals to set up plans depending on the chronological occurrence of activities they are involved in. Infeasibilities (such as unreasonable time requirements) can be detected and reported to the project leader who can then renegotiate commitments. For example, I-(3) is required for activities 3b-(3), 3b-(4) and 4-(2) simultaneously (in addition to other projects). If these requirements cannot be accommodated into I-(3)'s existing plan, the project leader can be informed automatically, so a reassignment can be made.
- When expectations about activities change, identify those parts of the project that must be restructured if necessary, and inform individuals about how their plans are affected. For example, when the software package purchased from outside could not be modified, activities 3a-(2), 3a-(3) and 3b-(2) were "cancelled", and the chronologically later activities cancelled. This required changes in information requirements (activity 1), which in turn led to restructured (renegotiated) applications development subprojects. The integrated tool could be used in the restructuring process just as it was used to set up the initial project schedules and arrangements. For example, if a project leader from either division requested that employees' time on this project be in-

creased and certain employees be reassigned, the integrated tool could determine the effect of the reassignment on the project's overall schedule and on individuals' plans.

Ambiguities in project priorities due to management "power struggles" cannot be dealt with by the problem-solving component of the integrated tool. However, the communications component does play an explicit role in negotiations about commitments through its structuring of requests and acceptances. If individuals fail to resolve conflicts with the available information about project activities and individual plans, the system will nevertheless provide a relatively objective record of the conflict which can be useful in addressing problems to management.

Conclusion

At the outset of this chapter, we argued that designers' ontological assumptions about collaboration influence the primitives of computer-based tools that are designed to support work groups.

We categorized support tools into two types, namely communications-oriented and problem-solving-oriented tools. We critiqued these tools by first identifying factors that shape collaborative work and the assumptions about collaboration that are implicit to these tools. The integrated view we then presented is an attempt at introducing the appropriate set of assumptions about collaboration into a support tool. From a pragmatic standpoint, this involves integrating the strengths of existing models in a way that allows project managers and individuals to deal more effectively with ambiguity, uncertainty, and complexity.

The general model proposed here has been motivated by learning experiences from other projects that we have been involved in. In one project (Dhar & Pople, 1987), we concentrated on building a primarily problem-solving-oriented system to support a problem in resource planning. Experience involving many individuals from different areas of an organization with this project has led us to believe that in order to enhance the utility of the problem-solver, a communications component would be useful. Similarly, evaluations of primarily communications-oriented tools (Dunham, Johnson, McGonagill, Olson, & Weaver, 1986) in field settings has led us to believe that tools alone do not capture the information necessary to enable individuals to resolve priority conflicts. As a next step in this research, we intend to define more precisely the structure of the general integrative model described in this chapter and the primitives that will be available to work group participants. We further intend to develop a prototype that can be a tool for empirical evaluations of work group collaborations under different conditions. This should enable us to understand more fully the details of how ambiguity, uncertainty, and complexity affect work group collaboration, and the scope of computer-based tools in dealing with these factors.

3

HOW IS WORK COORDINATED? IMPLICATIONS FOR COMPUTER-BASED SUPPORT

Bonnie Johnson
Corporate Technology Planning
Aetna Life & Casualty
Hartford CT

> In a certain sense we may say that man now has regained his former geocentric
> position in the universe. For a picture of the Earth has been made available
> from distant space and the sheer isolation of the Earth has become plain. This
> is as new and powerful an idea in history as any that has ever been born in
> man's consciousness . . . the planet Earth is a unique and precious place.
> —Anshen, 1973

INTRODUCTION

An inquiry in search of understanding coordination could be framed in many
contexts. One frame is that of "living on a small planet"; our competence at coordination is ultimately our salvation. The frame for discussing coordination in this
chapter is more mundane, perhaps, but not unrelated to this larger concern. The
frame here is "working in a large organization"; I discuss coordination technologies I promote as part of my job as a technology planner in Aetna Life &
Casualty's Corporate Technology Planning Unit.

CORPORATE TECHNOLOGY PLANNING AT AETNA

Corporate Technology Planning (CTP) is Aetna's advanced development
group. Our mission is to be a catalyst for introduction of new technologies
throughout the corporation. We help Aetna understand and use new technologies

in time to compete. Information systems, except for mainframe processing and telephone services, are largely decentralized to Aetna's six operating divisions. We are a corporate staff group with no authority to purchase production technologies. Eighteen professionals and four administrative personnel are responsible for identifying issues, researching business opportunities for specific technologies we assess are ready, and communicating with internal customers about the opportunities that these technologies might present. We succeed at these tasks when we help them use new technology in productive ways that were not possible before.

The task of engaging others in conversations about the potential of new technologies is difficult because people have little time for the future. Those in charge of developing and using computing systems make decisions on the basis of a set of immediate concerns, such as customer requirements and chargeback constraints, that are more real to them than speculations about future benefits of new tools. In order to identify issues and research opportunities, we have begun to operate through recognizing, and helping others to recognize, breakdowns in operations.

Breakdowns are moments of disruption or lack of fit, in which practices or attitudes that have been operating unnoticed receive attention. Breakdowns create opportunities by drawing attention to previously unnoticed possibilities. People do not stop to think about what they are doing until there is a breakdown.

With new systems technology, an announcement that a competitor has a new technology may lead to one's own computer system getting attention. The computer system that has been invisible suddenly becomes an object for evaluation; discussions about alternatives are welcome. A breakdown has occurred, not because there is something wrong with the computer system, but because now people are aware of its benefits and costs; they are now open to discuss alternatives and their relative advantages and disadvantages. Discussions of future computing architectures that seemed uninteresting because they were impractical only a few weeks before now become relevant and interesting. In general, then, our objective as a planning unit is to get existing ways of work noticed so that people will discuss with us alternatives to that way of working. These conversations are frequently informal, but they may also take the form of seminars and briefings we sponsor as well as occasional papers or reports we circulate.

UNDERSTANDING COORDINATION

At Aetna, as in most organizations, the role of coordination in daily business operations is not well understood. Most business analysts would not likely identify coordination as a key performance issue. However, many problems frequently cited by business analysts are symptoms of a widespread difficulty with coordination, such as:

- Bringing products to market too slowly
- Administrative red tape
- Excessive overhead expenses

Problems of coordination from which these apparent problems stem are typically too pervasive to be identified in a requirements analysis. The job of managers is coordination; but typically a manager is concerned with more apparent concerns—getting the product out, deciding who gets what bonus, managing an inventory to assure a steady flow of product. Moreover, managers are not quick to see that coordination problems are a likely remedy for their problems. For managers concerned with bringing products to market, computer tools for coordination may seem a diversion of precious computer resources. For managers frustrated with administrative red tape, computer tools that have to maintained, require special procedures, and change work habits may not seem prudent. For managers concerned with overhead, computer tools for coordination may seem to be just more overhead expenses.

Discussions of coordination could be framed in many ways. The focus of this chapter is coordination as a work issue; I, as a technology planner, examine and search for technologies to support it. My job function is to engage others in discussions about the possibilities offered by new technologies, finding champions to test these technologies, then evaluating these and, if appropriate, marketing them to others at Aetna.

In the following section I describe the traditional process of designing coordination tools and offer an alternative to the conventional vision. In the closing section I return to a discussion of CTP and provide examples of the general framework presented here in our initiatives to introduce coordination technologies.

APPROACHES TO DESIGN OF COORDINATION TOOLS

In this section, I describe two alternative approaches to understanding the issue of coordination for the purpose of design of tools to support it.

Designing from Artifacts and Residues

Users today have a choice of a wide variety of coordination technologies. Robert Johansen (chapter 1, this volume) characterizes these in detail. Inspecting these technologies, we might say that their designers were engaged in one or more of the following quests:

- *The perfect mail bag:* search for the perfect tool for non-simultaneous communication, often called "electronic mail";
- *The perfect calendar:* search for the perfect tool for scheduling one's body in proximity to others;
- *The perfect overhead projector:* search for the perfect tool for visual support in simultaneous communications;
- *The perfect telephone:* search for the perfect tool for bringing people together across distance.

In order to build tools, these designers analyze the artifacts and residues produced as humans coordinate their work. *Artifacts* of coordination are communications we produce in the effort to coordinate. These include internal and external correspondence, bills, invoices, and checks. *Residues* are the by-products of coordination or failures to coordinate such as telephone message slips. Implicit in the approach to design from artifacts and residues is the proposition that tools for coordination are tools for producing today's artifacts and residues. Hence, the term "electronic mail" makes sense to us today as a tool for coordination in the same way that the term "horseless carriage" made sense to people in 1915 as a tool for transportation.

Alternatively, consider the success of spreadsheet systems and whether they could have been created from analysis of artifacts. How many people would have said in 1975 that their work was fundamentally "spreadsheet" work? Those who looked to find residues of work on which to base new computer tools in 1975 would never have invented a spreadsheet program.

Designing for Conversations

We do not know what will be the new metaphor of coordinated activity, but we can speculate about essential human activities. Coordination is conversation: speaking and writing as action. Understanding speaking and writing as action rather than the artifacts of speaking and writing such as memos may prove to be a more fruitful avenue for understanding coordination and for building tools to support it. Design of tools for coordination may not proceed fruitfully through analysis of artifacts of conversations. Consider this conversation.

Mr. Bono, a mass observer, is making his rounds Tuesday morning, the day for ringing doorbells, assessing tithes, leaving notes for solicitors, and assisting those caught in chancery. A full day's work! Now he calls on Mrs. Dorfman, Apt. 1 in a modest duplex.
"Ring," says his finger.
"Yes?" replies Mrs. Dorfman.

Mr. Bono stands poised and open to all possibilities, his sensory gate ajar, his cognitive throughway green-lighted. "Yes," he notes aloud, "affirmative."

"If you're looking for Harry," Mrs. Dorfman says, "he's not here."

"Harry," says Mr. Bono, and smiles.

The delight of relevancies, the superb pattern of interconnections. A moment ago, Harry was a possibility; now Mr. Bono and Harry are joined in the infinite web of actuality. And Mrs. Dorfman too; a piece of the puzzle, a voice of reassurance and union in the silent interiors of streets and viaducts. O jack of selective trades! Artists of mass-construction!

"Mrs. Dorfman," says Mr. Bono, "I congratulate you and Harry too on your contribution to the What of things, the magnificence of mass."

"I'm sorry, we're not Catholics," Mrs. Dorfman replies.

Another line sent tingling into the net of circumstance, a bird for paradise, the innards of creation.

"We are all just men here," quotes Mr. Bono from recent reading.

"You mean the ecumenical thing?" asks Mrs. Dorfman.

"I refer to Leviathan," says Mr. Bono, "the vast summation whose integers we are whether hidden in homes or conversing as fresco."

"You must have the wrong house," Mrs. Dorfman says, "Nobody by that name lives here."

"And closes the door. Mr. Bono makes a notation in his blue notebook the color of Tuesdays, and turns toward the next house." (Natanson, 1970, p. 56-57).

This conversation is "nonsense." Because it "doesn't make sense," it reveals how conversations do make sense to us. With this imaginary conversation philosopher Maurice Natanson exemplifies sociality. "Observing the other means perceiving his action as already significant in terms of what that action means to him." This conversation reveals coherence to us precisely because coherence is missing. Mrs. Dorfman continually interprets Mr. Bono's words in ways to make them meaningful to her according to her interpretations of actions Mr. Bono could be intending. We understand that coherence is missing in that Mrs. Dorfman assesses she is unsuccessful in her actions to interpret Mr. Bono's words. The imaginary conversation exemplifies a breakdown in conversation. An alternative to design from examination of artifacts and residues is design through examination of breakdowns in conversation.

COORDINATION OF ACTION

I offer the following alternative view: Distinctions of human coordination can be abstracted from examination of conversation. By examining closely what people are doing when they coordinate their efforts, one can observe:

- Expression through subtle variations in sound and image,
- Coherence: each message is relevant to the preceding message,
- Interpretations and assessments constructed by people from their backgrounds and from a gestalt of sound and image messages; these are frequently noted orally or in writing,
- Enactment of routine behaviors; people do what is normal to do,
- Use of language to make, keep, and break commitments,
- Identifying what one wants to do from what one perceives as possible,
- Learning new interpretations and seeing new possibilities together with others.

Distinctions in Communication

To use these distinctions as a basis of design of computer tools to support coordinated actions, another set of distinctions is helpful: distinctions of communication.

Simultaneous communication (speaking to each other at the same time) has advantages and disadvantages. In simultaneous communications, people are more likely to invent new possibilities for themselves. The immediate feedback of the responses of the other promotes more consensual understandings.

Simultaneous communications can be ineffective and inefficient. In simultaneous communication people cannot edit their messages to "say what they mean," to respond appropriately to others, to present their messages in a forceful way, for example. Thus each may leave a simultaneous situation saying, "if only I had said ...." " Simultaneous communications are likely to be inefficient for each person because they must accommodate their individual schedules to be in the presence of the others.

The interpretation of written messages is a different experience than the interpretation of oral messages (Arnold, 1968). For example, because one can ponder and re-read them, written messages are interpreted more "rationally." In interpreting written messages, visual impact plays a role as well as verbal content. The richer the cues to interpretation, the more likely an interpreter will make a "sound" judgment (one that is "politically" astute, that is "socially" acceptable, that is sustained over a period of time).

Media choice is largely a taken-for-granted practice guided by what media are convenient, familiar, and deemed appropriate to the action.

Implications for Computer Support of Coordination

The most effective way of coordinating action is to put everyone in face-to-face contact (for example with desks in an open space, close to each other) and keep them that way. Some Japanese corporations put their top executives in such bullpen situations instead of the separated palaces more typical of American organizations.

Even this method has problems, however. First, of course, we cannot do global business this way. Even those in the Japanese bullpens must coordinate with people outside their view. Also, face-to-face communications, as we have observed, can be very ineffective ways of coordinating.

New problems arise when we attempt to coordinate across distance and time. Routine practices are disrupted; we see breakdowns more distinctly.

In examining computer tools to support human coordination across time and distance, the simplest set of requirements a user might pose is:

- Can I reach them?
- Do we have sufficient cues to be confident of our interpretations of each other? For example, do I have sufficient cues to know whether to trust the other; to know what is important to the other; to understand the other's background for understanding what I am saying?
- Do I have the means now for clarifying commitments and expectations?
- Do I have the means for managing commitments and expectations in the future?
- Is my tool transparent to me or do I have to stop to think about the tool every time I use it?
- Is my tool transparent to me when I pay for it because it is so essential to the work that I regard it as simply a "cost of doing business"?

A system for supporting coordination that could do all these things would likely be effective, but it might not be efficient. A system for guaranteeing that others are always available is likely to be very inconvenient at best. A system for guaranteeing sufficiency of cues is likely to barrage people with so many cues they cannot judge which is important. And so criteria of convenient access or normal-to-use are significant features for a coordination support system.

Computer support for coordinated work will consist of a set of technologies that supplement, and only in a limited sense, substitute for older systems of coordination. The significant possibility of these technologies is not their substitution, but rather the new ways of coordinating they can enable. They become widespread only when their use is so normal they become transparent.

The cost of coordination is transparent to most of us. We do not add up all the time we spend on telephone tag; we never combine the mail bill, the telephone bill, and so on. Researchers and implementors of office automation systems have attempted to justify the cost of equipment by proving the time we spend playing telephone tag and multiplying that times our wages. Such arguments often fall on deaf ears of executives who realize they are unlikely to reduce their staff by 5% even though each and every person saves a half an hour a day of telephone tag. The concept of transparency can be useful in understanding the total cost of a system. Not only must a system be transparent to use, its cost must be transparent. The telephone itself, and the company mail department, are examples. No one could imagine a modern organization without telephones or ways of distributing the mail. The cost is an essential part of doing business. The cost of any coordination system must be so apparently an essentially expenditure that one does not question whether money should be spent, although one may question the amount.

In summary, the perspective described here is that people coordinate actions with each other through our interpretation of gestalts of sound and image messages in a background of what they believe is already going on. In a program of computer support for coordinated work, therefore, key distinctions are sound, image, and background.

IMPLEMENTING COORDINATION TECHNOLOGIES AT AETNA

We can now examine the implications of coordination as conversation for assessing and implementing technology in a large organization such as Aetna. As noted in the opening section, CTP operates by observing and engaging others in discussions about the potential of new technologies. In order to get attention, we observe when people have breakdowns in which their normal ways of working become apparent and open to question. Four kinds of breakdowns create openings for discussions of coordination technologies: access, identity, process, and project.

Access

People notice that work is not being done well because of their inability to reach others. They become irritated with telephone tag.

With technology, it is possible to improve access of members of the organization to each other, and of external people such as customers to the appropriate people within the organization.

Telephone and mail service are obviously the most pervasive access technologies. But as people are increasingly unavailable by telephone, it becomes less effective. As costs for people and transportation increase, both internal and external mail service become less attractive.

Electronic mail, computer conferencing, and voice messaging are among the newer technologies that address access breakdowns. At Aetna, we determined that for the simple requirement of access, voice messaging is currently the most effective technology. Access technologies should be very inexpensive to use so everyone can be reached. Appropriate access technologies get messages to people easily, immediately, and cheaply. Access means having many cues to interpret what others are saying and to be able to converse, despite time and distance constraints, with considerable conversational coherence. People who want access do not necessarily require a record of the communication; ephemeral communications such as voice may be all that is required. Implementing effective access technologies requires providing support for users to learn very easy systems and teaching managers to take advantage of the opportunities the technologies provide. Voice messaging is most easily available to everyone because they already have the equipment on their desk, the telephone; they already know how to use most of the interface.

When we began to interest people at Aetna in voice messaging, we assessed that we were several years behind the competition. Our competitor across town had over 15,000 people already using voice messaging. We gave mailboxes to over 500 people, strategically located in units of the company with a direct bearing on our profitability. We encouraged them to use it and to talk about it with others. We found that those who used it as a coordination technology by sending messages to others on the system rather than just as an answering machine got substantially more benefits, over ten times a month more by our financial calculations.

Subjects in our voice messaging pilot were clear that voice messaging brings them two benefits: It is affordable and easy, but more important it provides them with more cues to understanding the other person than writing. The intentions of another person were more accessible to them when hearing the person's voice than from written messages or messages interpreted through intermediaries. We have an active program to promote voice messaging as a basic tool for improving access.

Identity

People notice that the memos, announcements, and other documents they send to others, inside the company and outside, do not represent them well. The documents are ineffective at presenting messages in an appealing yet concise style. Furthermore, using new technologies, people represent the company in unauthorized ways, hence putting the corporate identity at risk.

With technology, it is possible to improve the effectiveness of documents and other presentation material to create favorable images of the sender and to improve the speed and effectiveness of interpreting the message through use of graphic design.

The phrase "a picture is worth a thousand words" can explain why many are interested in identity technologies. Now almost anyone, without expensive printing equipment or special talent, can produce graphic material. As individuals and as corporations, we seek tools that represent us well to others. With graphics tools one can create a one-page message that has the same visual impact that otherwise might have required several pages of text.

We have found that tools that promote identity by improving the appearance and effectiveness of documents cannot be strictly controlled. With desktop publishing people create documents without centralized quality control. The technology is expensive enough that no one regulates its use. People will find ways to acquire and use these technologies because they recognize that they promote a better way of doing business for them. We have a proliferation of desktop publishing systems at Aetna, each doing a critical job of communicating. Our old notions of graphics and printing are obsolete. Graphic tools in themselves are not enough for inventing our public identities. Use of these tools to enhance coordination requires competence in creating an effective look and content. We have new options for choosing tools to meet timeliness and cost criteria. We have a program in desktop publishing aimed at developing and teaching new methods for making the most effective use of graphics and desktop publishing tools for improving coordination. We participate in a corporate identity program examining and improving how we present ourselves through communication.

I have discussed voice messaging as an *access* technology and graphics/desktop publishing as an *identity* technology. However, we also find that through desktop publishing, we are able to access more people; we are designing our use of voice messaging, particularly as the messaging system answers calls from customers, to present an effective identity for the company. The point here is not to isolate particular technologies as one or the other, but to point to the breakdowns of access and identity that constitute coordination, and how technologies that provide rich clues to interpretation can enhance these.

Process

People notice that processing work using paper files is slower than electronic processing. After three decades of automation, much of our production work consists of passing paper files. People are annoyed that only one person can use a file at the same time; others must rearrange their work while waiting.

With technology, it is possible to improve sequential work-flow management by using electronics to ship, store, and manage the flow of electronic images of pages that replace the original documents.

In insurance, we produce promises on paper. Requests for action, such as requests to issue a policy or to pay a claim, come mainly on paper; the results of what we do go to customers on paper in the form of a legal agreement or a check. Much of the coordination is not only paper-bound but sequential. My group is investigating and promoting document image processing as a technology to address sequential coordination. Technologies for coordination in a process dependent on external paper have not changed much since the development of the vertical file. Micrographics, for example, provides a compact way of storing paper, but made sequential coordination more difficult because filming made papers difficult to read; it is therefore inappropriate as a coordination technology.

New technologies such as write-once optical storage now open up new options for redesigning sequential work flows and vastly improving coordination. Cheaper storage makes it feasible to use electronics to manipulate the image of a paper; thus, a user can have the clues to interpretation available on the original document and yet be able to manage the flow of work electronically. We are experimenting with implementing such technologies through understanding of the work to be done and joint evolution of the social as well as the technical process. Our first experiment is with a mortgage loan application where files contain legal documents from outside the company. In other applications we will be replacing claim applications and underwriting files. Several image processing vendors, notably Filenet, Kodak, and Wang, are developing sequential coordination tools in the form of work-flow software for managing the movements of images to different people. We are also finding it possible to write our own software to design sequential flow of document images.

Project

People notice that they and others slip deadlines, miss meetings, and have difficulties managing to accomplish several priority tasks. In boxes become full of paper, often requiring managers to hire others to read and respond to the requests in those boxes.

With technology, it is possible to improve reciprocal work flow by using electronic communications to track requests, deadlines, meeting commitments, and other priorities. Project coordination is sensitive to commitments and time frames. In sequential work flows, jobs can be designed to secure cooperation that is enforced through supervision. In project coordination, jobs are more loosely coupled, and tasks evolve throughout the project; management may be matrixed so that responsibility and authority are unclear. Control of project accomplishments depends on the ability of people to secure commitments and track action in a timely manner without depending on stable procedures or authority.

Most project management tools are planning and tracking tools, rather than coordination tools per se. That is, they allow planners to specify what needs to be done and then record what has been done, recalculating time projections, and so on. A new class of "intelligent" conferencing tools, such as The Coordinator Workgroup System by Action Technologies, Inc. supports reciprocal coordination by tracking requests, deadlines, and priorities. Users declare the intent of their messages, for example to make a request of another for action to be done by some deadline, or to open a discussion of what might be done at some unspecified date. Users of The Coordinator, unlike Mr. Bono quoted earlier, state their intent to aid the reader's understanding of the electronic message and also to make the messages briefer. The Coordinator's "conversation manager" tool allows the reader to trace back from any message to read the history of the conversation (the previous messages within a line of conversation), and thus to remind themselves of the background of the current message.

In 1986 we experimented with The Coordinator as a tool for improving project coordination and found that it sped the accomplishment of tasks by helping to clarify what was to be done and allowing people to be reminded of this history of issues as new and related issues came up. We also found that many potential users were reluctant to learn a new language, questioning whether their coordination could actually be made more explicit in their communications. We are currently implementing a new experiment with a systems development team to examine the impact of language changes on project productivity.

CONCLUSION

This chapter is an introductory exploration of some distinctions of coordination that have implications for computer support in general and are guides to assessing and introducing new technologies to improve coordination at Aetna. To develop effective computer support for coordination, designers should support conversations through tools that provide rich sound and image cues and rich backgrounds for interpreting messages.

As a background for discussing coordination, I observed that my own work as a technology planner at Aetna is done through conversations. I described how we in Corporate Technology Planning have learned to be effective in innovating with new technology from a staff position. Heidegger's concept of breakdown is fundamental to our work. We can engage others in discussions about the future possibilities that new technologies represent to the extent that they experience breakdowns that call attention to current practices and technologies.

I compared coordination as conversation to the more traditional approach to designing coordination technologies from artifacts and residues. Products such as electronic mail support the results of our attempts to coordinate mail, an artifact, rather than address the objective: to improve coordination. I speculated that effective tools of coordination will support breakdowns in conversation. I then used an imaginary conversation to illustrate some principles of conversation and speculate on some breakdowns. I distinguished some features of effective conversations.

Finally, to illustrate the concept of coordination as conversation, I returned to our work at Aetna, identifying four breakdowns of coordination that we are addressing with computer tools: access, identity, process, and project. I described tools for managing these breakdowns including voice messaging, desktop publishing, image processing, and intelligent conferencing. These tools provide rich cues to interpretation, promote coherence among messages in a conversation, and are normal to use. As such they promote coordination by improving our capability to converse with each other.

ACKNOWLEDGMENTS

My description of coordination, communication, breakdown, and conversation owes much to the published and unpublished papers of Fernando Flores and his associates (Flores & Bell, 1984; Flores & Ludlow, 1981; Winograd & Flores, 1986). I do not claim that my analysis is Flores' or represents his well; only that the analysis I present here is based in my interpretation of his distinctions.

I also acknowledge Daniel Shurman of Humanware who taught me and others in my work group to use the distinctions of the Coordinator Workgroup System and thereby improved our effectiveness and our understanding of what is possible with coordination technologies.

Jeffry Alperin, Assistance Vice President in charge of CTP, uses language-for-action to create a context for our work and shows me the pragmatic value of the perspective on coordination presented here. My colleagues, Geraldine Weaver, Amy Frink, Pauline Miller-Jackson, and Betsy Shaw constantly coach me in the pragmatic distinctions of communication and interpretation.

Several research colleagues including Paul Saffo, William Nothstine, Margi Olson, Bob Dunham, Don Van Doren, Ron Rice and Grady McGonagill have consulted with us in research projects that have expanded our outlook.

4

THE INFORMATION LENS: AN INTELLIGENT SYSTEM FOR INFORMATION SHARING AND COORDINATION

Thomas W. Malone, Kenneth R. Grant, Kum-Yew Lai, Ramana Rao, and David A. Rosenblitt
Massachusetts Institute of Technology

INTRODUCTION

One of the key problems when any group of people cooperates to solve problems or make decisions is how to share information. Thus one of the central goals of designing good "organizational interfaces" (Malone, 1985) should be to help people share information in groups and organizations. In this chapter, we describe a prototype system, called the Information Lens, that focuses on one aspect of this problem: how to help people share the many diverse kinds of qualitative information that are communicated via electronic messaging systems. We also show how the same general capabilities that help with information sharing can be used to support a variety of more specific coordination processes such as task tracking and meeting scheduling.

It is already a common experience in mature computer-based messaging communities for people to feel flooded with large quantities of electronic "junk mail" (Denning, 1982; Hiltz & Turoff, 1985; Palme, 1984; Wilson, Maude, Marshall, & Heaton, 1984), and the widespread availability of inexpensive communication capability has the potential to overwhelm people with even more messages that are of little or no value to them. At the same time, it is also a common experience for people to be ignorant of facts that would facilitate their work and that are known elsewhere in their organization. The system we describe helps solve both these problems: It helps people filter, sort, and prioritize messages that are already addressed to them, and it also helps them find useful messages they would not otherwise have received. In some cases, the system can respond automatically to certain messages, and in other cases it can suggest likely actions for human users to take.

The most common previous approach to structuring information sharing in electronic messaging environments is to let users implicitly specify their general areas of interest by associating themselves with centralized distribution lists or conference topics related to particular subjects (e.g., Hiltz & Turoff, 1978). Because these methods of disseminating information are often targeted for relatively large audiences, however, it is usually impossible for all the information distributed to be of interest to all recipients.

The Information Lens system uses much more detailed representations of message contents and receivers' interests to provide more sophisticated filtering possibilities. One of the key ideas behind this system is that many of the unsolved problems of natural language understanding can be avoided by using semi-structured templates (or frames) for different types of messages. These templates are used by the senders of messages to facilitate composing messages in the first place. Then, the same templates are used by the receivers of messages to facilitate constructing a set of rules to be used for filtering, categorizing, and otherwise processing messages of different types.

Examples

Before describing the Information Lens system itself, we briefly describe a few examples of situations in which systems like this might be useful. Later, we discuss in more detail how the capabilities of the Lens system could help in these situations.

Distributing engineering change notices. Almost all organizations that design and manufacture complex physical products (such as cars, televisions, or computers) have some system for distributing information about changes in design specifications. The forms that contain this information are often called "engineering change notices" or "engineering change orders." A particular engineering change may be of interest to a variety of people including (1) design engineers working on related parts of the product, (2) manufacturing engineers designing the manufacturing process, and (3) procurement specialists who must purchase the component parts to be used in the product. Early in the design process, an engineering change may be made by an individual engineer. As the design progresses, approval from group or division managers might be necessary. After the product is actually in production, some engineering changes may require approvals from very senior executives such as vice presidents.

How should this process be managed? How does the engineer who originates a change know who else will be interested? How can engineers designing one part of the product know that they have seen all the changes to other parts that might affect their work? What happens when incompatible changes are not detected until the product is actually in production? Many organizations today still use a primarily paper-based system to manage this process and they rely on official organizational channels and informal personal networks to disseminate information. We see later how systems like the Information Lens may dramatically improve the speed and effectiveness of this process.

Handling software bug reports. In any multi-person software development project, there must be some way of handling bug reports from users and testers. For example, who should screen out or respond to incomprehensible or erroneous bug reports? How is a bug report distributed to developers who should know about it or who might have information related to fixing it? Who insures that someone takes responsibility for fixing legitimate bugs? How are others notified when a bug is fixed? In many organizations today, this process relies heavily on communication by electronic mail, but we see later how a system like the Information Lens could help manage the process more effectively.

Distributing news and rumors to traders in financial institutions. In modern financial markets, perhaps as much as anywhere else in the world, "time is money"; that is, differences of a few minutes in getting access to the right information can make the difference between gains or losses of millions of dollars. How can traders find out quickly about news stories or rumors that will affect their markets without being overwhelmed by far too much irrelevant information? Athough there are already various forms of market quotation and news wire services used to support traders in these environments, we see later how ideas like those incorporated in the Information Lens might be even more helpful.

Sharing "lore" about repair problems in field service groups. Many manufacturing companies have widely distributed networks of field service offices that are responsible for maintaining and repairing their products for customers. Even though, in some ways, each customer's site and product is unique, there are many commonalities in the kinds of problems that arise at different customer sites and at different field service offices. We see later how systems like the Information Lens could help repair personnel share their experiences with different kinds of problems without overloading each other with too much irrelevant information. Such systems could also help people quickly find the company-wide experts on particular kinds of problems or the answers to specific questions.

SYSTEM DESCRIPTION

There are five key ideas that, together, form the basis of the Information Lens system. Though some of these ideas are empirically testable hypotheses, we treat them here as premises for our system design. We list and briefly describe these ideas here. In the next sections, we describe in more detail how the Lens system uses them:

(1) A rich set of semi-structured message types (or frames) can form the basis for an intelligent information sharing system. For example, meeting announcements can be structured as templates that include fields for "date," "time," "place," "organizer," and "topic," as well as any additional unstructured information. There are three reasons why this idea is important:

(a) Semi-structured messages enable computers to automatically process a much wider range of information than would otherwise be possible. By letting people compose messages that already have much of their essential information structured in fields, we eliminate the need for any kind of automatic parsing or understanding of free text messages while still representing enough information to allow quite sophisticated rules to process the messages.

(b) Much of the processing people already do with the information they receive reflects a set of semi-structured message types. In our informal studies of information sharing in organizations (Malone, Grant, Turbak, Brobst, & Cohen, 1987), we found that people often described their filtering heuristics according to categories of documents being filtered (e.g., "This is a brochure advertising a seminar. I usually throw these away unless the title intrigues me or unless it looks like a brochure I could use as a model for the ones I write.")

(c) Even if no automatic processing of messages were involved, providing a set of semi-structured message templates to the authors of messages would often be helpful. Two of the people in our informal interviews mentioned simple examples of this phenomenon: one remarked about how helpful it would be if any memo requesting some kind of action included, in a prominent place, the deadline by which the action needed to be taken; a second commented about how wonderful it would be if all the meeting invitations he received included a field about why he was supposed to be there. We see later how message templates can be provided in a flexible way that encourages, but does not require, their use.

(2) Sets of production rules (that may include multiple levels of reasoning, not just Boolean selection criteria) can be used to conveniently specify automatic processing for these messages.

(3) The use of semi-structured message types and automatic rules for processing them can be greatly simplified by a consistent set of display-oriented editors for composing messages, constructing rules, and defining new message templates.

(4) The definition and use of semi-structured messages and processing rules are simplified if the message types are arranged in a frame inheritance lattice.

(5) The initial introduction and later evolution of a group communication system can be much easier if the process can occur as a series of small changes, each of which has the following properties: (a) individual users can continue to use their existing system with no change if they so desire, (b) individual users who make small changes receive some immediate benefit, and (c) groups of users who adopt the changes receive additional benefits beyond the individual benefits.

System Overview

In order to provide a natural integration of this system with the capabilities that people already use, our system is built on top of an existing electronic mail system. Users can continue to send and receive their mail as usual, including using centrally maintained distribution lists and manually classifying messages into folders. In addition, the Lens system provides four important optional capabilities: (1) People can use structured message templates to help them compose their messages; (2) Receivers can specify rules to automatically filter and classify messages arriving in their mailbox; (3) Senders can include as an addressee of a message, in addition to specific individuals or distribution lists, a special mailbox (currently named "Anyone") to indicate that the sender is willing to have this message automatically redistributed to anyone else who might be interested; and (4) Receivers can specify rules that find and show messages addressed to "Anyone" that the receiver would not otherwise have seen.
By gradually adding new message types and new rules, users can continually increase the helpfulness of the system without ever being dependent on its ability to perfectly filter all messages.

System architecture. The Lens system was written in the Interlisp-D programming environment using Loops, an object-oriented extension of Lisp. The system runs on Xerox 1100 series processors connected by an Ethernet. We use parts of the Lafite mail system and the XNS network protocols already provided in that environment. The message construction aids and the individual filtering rules all operate on the users' personal workstations.

As Figure 4.1 illustrates, messages that include "Anyone" as an addressee will be delivered by the existing mail server directly to the explicit addressees as well as to an automatic mail sorter that runs on a workstation and periodically retrieves messages from the special mailbox. This automatic mail sorter may then, in turn, send the message to several additional recipients whose rules selected it.

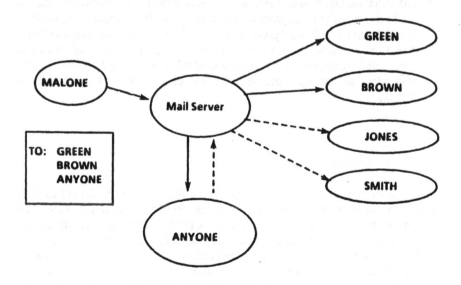

Figure 4.1. The Lens system includes components in the users' workstations and in a central server (called "Anyone"). Messages that include "Anyone" as an addressee are automatically distributed to all receivers whose interest profiles select the messages as well as to the other explicit addressees.

Implementation status. The Information Lens system as described in this chapter currently exists in prototype form. As of this writing, the system has been in regular use by about five members of our research group for over a year, and a larger scale test at an industrial research center has begun.

Messages

The Lens system is based on a set of semi-structured messages. For each message type, the system includes a template with a number of fields or slots for holding information. Associated with each field are several properties, including the default value of the field, a list of likely alternative values for the field, and an explanation of why the field is part of the template.

Figures 4.2 and 4.3 show a sample of the highly graphical interaction through which users can construct messages using these templates (see Tou, Williams, Fikes, Henderson, & Malone, 1982, for a similar approach to constructing database retrieval queries). After selecting a field of a message by pointing with a mouse, the user can point with the mouse again to see the field's default value, an explanation of the field's purpose, or a list of likely alternatives for filling in the field. If the user selects one of these alternatives, that value is automatically inserted in the message text. The user can also edit any fields directly at any time using the built-in display-oriented text editor. For example, the user can add as much free text as desired in the text field of the message.

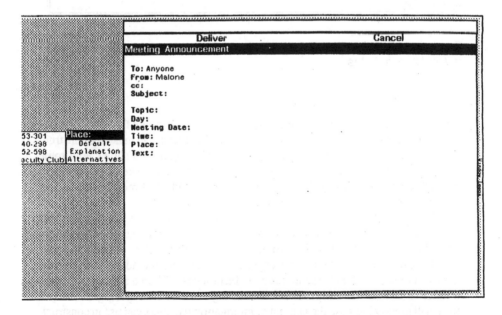

Figure 4.2. Messages are composed with a display-oriented editor and templates that have pop-up menus associated with the template fields.

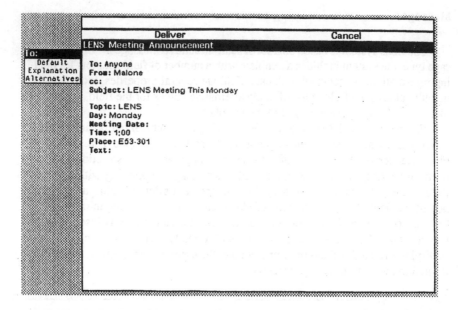

Figure 4.3. Some templates already have a number of default values filled in for different fields.

By providing a wealth of domain-specific knowledge about the default and alternative values for particular types of messages, the system can make the construction of some messages much easier. For example, Figure 4.3 shows how some message templates, such as a regular weekly meeting announcement, may have default values already filled in for most of their fields and require only a few keystrokes or mouse clicks to complete and send off.

Users who do not want to take advantage of these message construction aids can simply select the most general message type (message) and use the text editor to fill in the standard fields (To, From, and Subject) just as they would have done in the previous mail system. We expect, however, that the added convenience provided to the senders by semi-structured templates will be a significant incentive for senders to use templates in constructing some of their messages. This, in turn, will greatly increase the amount of information receivers can use in constructing automatic processing rules for incoming messages.

Message Types

To further simplify the construction and use of message templates the templates are arranged in a network so that all subtypes of a given template inherit the field names and property values (e.g., defaults, explanations, and alternatives) from the parent template. Any subtype may, in turn, add new fields or override any of the property values inherited from the parent (e.g., see Fikes & Kehler, 1985). For example, the seminar announcement template adds a field for speaker that is not present in its parent template meeting announcement. The Lens meeting announcement (Figure 4.3) adds a number of default values that are not present in its parent. The inheritance network eliminates the need to continually re-enter redundant information when constructing new templates that resemble old ones, and it provides a natural way of organizing templates, thus making it easier for senders to select the right template.

The message type lattice is made visible to the user through the message type browser. Figure 4.4 shows this lattice browser for our sample network of message types. Users select a template to use in constructing a new message by clicking with the mouse on the desired message type in this browser. By clicking with a different mouse button, users can view or modify the rules (see below) associated with a particular message type. Like the other message type characteristics, these rules are inherited by the subtypes of a message template. Thus, for example with the network shown in Figure 4.4, the rules for processing "notices" and "messages"

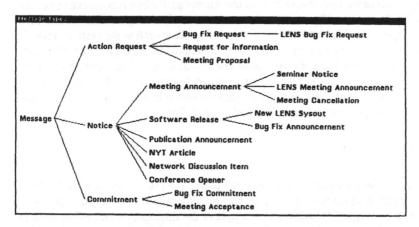

Figure 4.4. The message templates are arranged in a network with more general types at the "top" (shown at the left) and more specific types shown at the "bottom" (shown at the right).

would be applied to incoming "meeting announcements" as well as the rules specifically designed for meeting announcements.

The network shown in Figure 4.4 includes some message types that we believe will be useful in almost all organizations (e.g., meeting announcements) and some that are important only in our environment (e.g., Lens meeting announcement). Different groups can develop detailed structures to represent the information of specific concern to them.

We have developed another display-oriented editor, like the message editor shown in Figures 4.2 and 4.3, for creating and modifying the template definitions themselves. We expect that in some (e.g., rarely used) regions of the network anyone should be able to use this "template editor" to modify an existing message type or define a new one, whereas in other regions, only specifically designated people should have access to this capability. In the current version of the system people can use a simple version of this editor to personalize the default, explanation, and alternatives properties of the fields in existing message types.

Group use of message types. Individuals who begin using this system before most other people do can get some immediate benefit from constructing rules using only the fields present in all messages (To, From, Subject, Date). Groups of individuals who begin to use a set of common message types can get much greater benefits from constructing more sophisticated rules for dealing with more specialized message types. For example, a general rule might try to recognize "bug reports" based on the word "bug" in the subject field, but this would be a very fallible test. A community that uses a common template for bug reports can construct rules that deal only with messages the senders classify as bug reports. These rules can use specialized information present in the template such as the system in which the bug occurred, the urgency of the request for repair, and so forth. From the viewpoint of organization theory, we know that "internal codes" are among the most important productive assets of an organization (Arrow, 1974; March & Simon, 1958). In effect, the Lens system provides a medium in which this collective language of an organization can be defined and redefined.

Rules

Just as the structure of messages simplifies the process of composing messages, it also simplifies the process of constructing rules for processing messages. For instance, Figure 4.5 shows an example of the display-oriented editor used to construct rules in the Information Lens system. This editor uses rule templates that are based on the same message types as those used for message construction, and it uses a similar interaction style with menus available for defaults, alternatives, and explanations. We expect that this template-based graphical rule construction will be much easier for inexperienced computer users than more conventional rule or query languages. For example, the users of most typical database retrieval systems must already know the structure of the database fields and their plausible values in order to construct queries. Users of our system have all this information immediately available and integrated into the rule-construction tools (Tou et al., 1982).

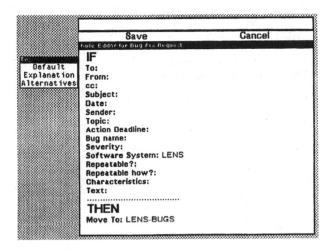

Figure 4.5. Rules for processing messages are composed using the same kind of editor and the same templates as those used for composing messages in the first place.

Local rules. There are currently two kinds of rules in the Lens system: "local rules" and "central rules". Local rules are applied when messages are retrieved from the mail server to a user's local workstation. Typical rule actions possible here include: moving messages to specific folders (Figure 4.6a), deleting messages (Figure 4.6b), and automatically "resending" messages to someone else (Figure 4.6c). Both "move" and "delete" mark a message as deleted, but do not physically expunge it. Thus, subsequent rules can move different copies of the same message to different folders. "Resending" a message is similar to "forwarding" it, except that instead of copying the entire original message into the body of a new message, the new message preserves the type and all but the "To" and "cc" fields of the original message. The new value of the "To" field is a parameter the user specifies for the automatic rule action, and the user whose rule does the resending is added as the "Sender" of the new message.

(a) **IF**   Message type: Action request
             Action deadline: Today, Tomorrow
   THEN *Move to*: Urgent

(b) **IF**   Message type: Meeting announcement
             Day: Not Tuesday
   THEN *Delete*

(c) **IF**   Message type: Meeting proposal
             Sender: Not Axsom
   THEN *Resend:* Axsom

(d) **IF**   From: Silk, Siegel
      THEN *Set Characteristic:* VIP

      IF   Message type: Action request
             Characteristics: VIP
      THEN *Move to* : Urgent

(e) **IF**   Message type: Request for information
             Subject: Al, Lisp
   THEN *Show*

(f) **IF**   Message type: NYT Article
             Subject: Computer
   THEN *Move to*: Computers

      IF   Message type: NYT Article
             Subject: Movies
      THEN *Move to:* Movies

      IF   Message type: NYT Article
             Article date: < Today
             Characteristics: Not MOVED
      THEN *Delete*

      IF   Message type: NYT Article
             Characteristics: Not MOVED and Not DELETED
      THEN *Move to:* NYT Articles

Figure 4.6. Sample Rules.

When the local rules have finished processing all incoming messages, the numbers of new messages that have been automatically moved into different folders because the last time the folder was viewed are shown on a hierarchical display of the folder names. Messages that were not moved anywhere remain in the "root" folder.

By combining conditions within and between fields, users can construct arbitrary Boolean queries. More interestingly, users can also construct elaborate multi-step reasoning chains by having some rules set characteristics of messages and then having other rules test these characteristics (Figure 4.6d; see Malone, Grant, Turbak, Brobst, & Cohen, 1987, for details). It is also possible to have different kinds of actions available for different kinds of message. For example, rules for "meeting proposal" messages have an automatic action available to "accept" a meeting.

Central rules. In addition to the local rules applied when messages are retrieved to a user's workstation, an individual user can also specify central rules to select messages addressed to "anyone" that the user wants to see. Only two kinds of rule action are possible for central rules: "show" (Figure 4.6e) and "set characteristic". "Show" causes a message to be sent to the user, unless the user would have already received it as one of the original recipients. When these messages arrive at the user's workstation they are processed, along with all the other messages, by the user's local rules.

Rule interactions. Our early experience with the system suggested the importance of being able to give users some control over the interactions between local rules (e.g., having certain rules fire only if no other rules have fired on a message) and of being able to explain to users why certain messages were processed as they were. Accordingly, we added several simple features to the rule system. First, the rules for a given message type are applied in the order they appear in the "rule set editor" for that message type, and the rules pertaining to a specific message type are always applied before the rules inherited from more general message types. Also, rules that take actions (such as moving or deleting a message), always set a characteristic of the message (such as "MOVED" or "DELETED"). Thus subsequent rules can include conditions such as "the message has not yet been moved or deleted" (see Figure 4.6f). Finally, in order to help users understand and modify their rules, a simple explanation capability allows users to see a history of the rules that fired on a given message.

Intelligent Suggestions for Responding to Messages

The presence of recognizable types of semi-structured messages also simplifies the task of having the system intelligently present options for what a user might want to do after seeing a message. Almost all electronic mail systems provide standard actions (such as "answer" and "forward") that can be taken after seeing a message. We have recently generalized this capability in two ways.

Suggested reply types. First, since either answering or forwarding a message creates a new message, our system suggests options for the type of new message to be created. For instance, when a user selects the "answer" option for a "bug fix request", a pop-up menu appears with three choices: "bug fix commitment", "request for information", and "other". Selecting commitment results in a new "commitment" message being constructed to answer the message. Figure 4.7 shows a selection of message types and their default reply types.

Original message	Suggested reply types
Message	Message
Action request	Commitment, Request for information,
Action request, Message	
Notice	Request for information, Action request,
Message	
Bug fix request	Bug fix commitment, Bug fix announcement,
Request for information	
Meeting proposal	Meeting proposal, Meeting acceptance
Meeting announcement	Request for information

Figure 4.7 Sample of Message Types Automatically Suggested as Replies

Suggested response actions. Even more important than suggesting reply types, our system is also able to suggest other actions a user might want to take after seeing a message of a given type. For example, when a user reads a message of type "meeting announcement" another option, "add to calendar" is automatically presented, in addition to the standard options like "answer" and "forward." If the user selects this option, the information already present in the message in structured form (e.g., date, time, and meeting topic) are used to add an appropriate entry to the user's on-line calendar. As another example, when a user reads a message of type "software release" (or any of its subtypes such as "bug fix announcement"), an option called "load file" is automatically presented, and if this option is chosen, the file specified in the message is automatically loaded into the user's system. This saves the user from having to type the "load" command and the (often long) file designation.

APPLICATIONS

Our first, and most fully developed, application of these ideas is the intelligent information sharing system described above. In order to demonstrate the generality of the ideas, we now describe several other simple applications we have implemented in the same framework. The most important point here is not that it is possible to implement these other applications—in fact, we have chosen applications that have already been implemented in other systems. Instead the point is that a wide variety of applications for supporting cooperative work can all be implemented in a way that is (1) smoothly integrated from the user's point of view, and (2) much easier to implement because it takes advantage of the basic capabilities for supporting structured messages.

Computer conferencing. Many of the capabilities of a computer conferencing system (e.g., Hiltz & Turoff, 1978), can be easily incorporated in the Information Lens system. We have done this by adding (1) a new message type called "conference opener" that includes the name of the conference and (optionally) its parent conference, and (2) a response option called "join" for this type of message that, if chosen, will automatically create (a) a new folder by this name, (b) a new rule to select messages addressed to "Anyone" that contain this name in the topic field, and (c) a new rule to move all messages received about this topic to the new folder.

Our system thus includes the capability that computer conferencing systems have for structuring communication on the basis of flexibly defined sets of topics and subtopics. It is also easy to see how our system goes beyond these customary capabilities of a computer conferencing system by allowing, for example, more sophisticated rules that filter not only on topic but also on other characteristics such as sender.

The primary computer conferencing capability that is not included in our system so far is the ability for a user to retrieve messages that were sent before that user joined the conference. This capability is clearly important, and it would be quite desirable to add a shared database to our system. The addition of such a shared database of semi-structured messages could be done within the same general framework and would make possible much more sophisticated retrieval possibilities than are possible with only unstructured messages (e.g., see Tou et al.,1982).

Calendar management. We have already seen how a very simple form of calendar management is included in the system by providing users with a response option that will automatically insert incoming meeting announcements into the user's on-line calendar. We have also implemented a more sophisticated protocol for semi-automated meeting scheduling. This protocol uses several new message types, including "meeting proposals" and "meeting acceptances." Proposals and acceptances both include all the fields that "meeting announcements" do, but the values may often be nonspecific (e.g., "sometime this week" for the date field). People can schedule meetings by sending a sequence of proposals (and possibly counterproposals) until a proposal is accepted.

Our system provides automated support for this process in several ways. First, some people may want to automatically resend all messages of certain types (e.g., meeting proposals) to other people (e.g., their secretaries) for a response. Second, the system helps people construct replies to these messages. For instance, users can choose to reply to a "meeting proposal" with a "meeting acceptance" and all the information (such as time, place, and topic) will be automatically copied from the proposal to the acceptance. When a meeting acceptance is received, one of the action options presented is to "confirm" the meeting. Selecting this option automatically adds the meeting to the user's calendar and sends a meeting announcement to the other participant(s) confirming the scheduled time.

Although we have not done so yet, this general framework also makes it possible to have some meetings scheduled completely automatically (e.g., see Greif, 1982). For example, meeting proposals from certain people (e.g., members of one's own work group) might be automatically accepted if they fall within regular working hours and do not conflict with other meetings already scheduled. Any messages that are exceptional (e.g., a request to meet outside of regular working hours or a request to meet at a time when a conflicting meeting is already scheduled) will then be brought to the attention of the human user for special handling.

These systems are, by no means, a complete solution to the meeting scheduling problem. For instance, because people can find out about each other's schedules only through messages, the system may require a number of iterations just to eliminate times in which they have conflicts due to publicly scheduled meetings. Adding shared databases might help solve this problem, but it brings up other problems about who has access to which parts of other people's calendars. Even in a system with partially shared databases, the semi-structured approach we have described has the desirable property that some cases (e.g., some meeting requests) are handled automatically while others are handled by human users in a smoothly integrated way. We believe this "graceful degradation" property will be especially important to the acceptance of systems—like those for calendar management—that involve subtle interpersonal and political issues about which people are reluctant or unable to be explicit.

Project management and task tracking. Systems based on structured messages can support project management and other coordination processes by helping to keep track of what tasks have been assigned to whom (e.g., see Sathi, Fox, & Greenberg, 1985; Sluizer & Cashman, 1985; Winograd & Flores, 1986). One simple way for users to do this in the current system is simply to set up rules that move copies of all action requests and commitments into special folders. For example, action requests a user receives might be categorized in folders by the project to which they relate, whereas action requests a user sends might be categorized by the people to whom they are sent.

To illustrate how more elaborate capabilities can be built up within the same framework, we have implemented a simple task tracking system for software maintenance activities similar to the example described by Sluizer and Cashman (1985). In this application, "users" of a software system send "problem reports" to a "work assigner." The work assigner first sends an "acknowledgment" to the user and then sends the problem report to a "developer." When the developer fixes the problem, the developer sends the "fix report" to the work assigner who, in turn, sends a "user report" to the original user noting that the problem has been fixed.

This application was implemented in our system by defining the three message types (two of which already existed under different names) and adding several new response options. Users who play the role of work assigners have two possible response actions suggested when they view problem reports. If the problem report has not been acknowledged, the option presented is to acknowledge it. Selecting this option prepares a standard acknowledgment message with the appropriate fields filled in from the information on the original problem report. When the acknowledgment message is sent, the token "acknowledged" is added to the "characteristics" field in the problem report message. Whenever a problem report that includes "acknowledged" in its "characteristics" field is displayed, a response action of "assign" is suggested. Selecting this option prepares the problem report message for forwarding to a developer whose name the work assigner selects from a list of alternatives.

Clearly this task-tracking system is quite limited. The original XCP system, for example, used a set of primitive actions to define protocols involving a number of roles, message types, and actions. A similar capability would be desirable in our system. Another obviously desirable capability would be more elaborate database facilities for sorting, displaying, and modifying the status of tasks. For example, the current system can sort messages into folders according to various criteria and it can display the header information (e.g., date, sender, and subject) for the messages in a folder. But it would also be quite useful to be able to display the tasks one had committed to do in a report format, sorted by due date, that summarized the task names and the task requestors. If these database capabilities are implemented in a general way for structured objects, then they should be useful for developing many other coordination-supporting applications as well.

Connections with external information sources. One of the desirable aspects of the system architecture we have described is its versatility in dealing with external information sources. As previously indicated, messages that are sent to users of the Lens system from people who do not use the system are simply repesented by Lens as messages of type "message." When Lens users send messages to others outside the system, all the fields in the template that are not part of the standard message header are sent as part of the text field.

It is also possible to do more intelligent translation into and out of certain message types. For instance, we currently receive daily on-line transmission of selected articles from the New York Times (via the system developed by Gifford, Baldwin, Berlin, & Lucassen ,1985). When the Anyone server receives these messages, it parses out the fields already present in the wire service feed (e.g., "title", "subject", "category", and "priority") and formats these messages as Lens templates with the same fields. Users of our system are then able to construct elaborate sets of production rules to select the articles they wish to see and sort them into categories.

One of the most interesting possibilities for such systems occurs in the formation of computer-based markets (e.g., see Malone, Yates, & Benjamin, 1987). For example, techniques like those described here can be used to screen advertising messages and product descriptions according to precisely the criteria that are important to a given buyer. Air travellers, for instance, might specify rules with which their own "automated buyers' agents" could examine a wide range of possible flights and select the ones that best match that particular traveller's preferences. The preferences might include decision rules for trading off between factors such as cost, convenient arrival and departure times, window seats versus aisle seats, minimum number of stops, and so forth. A fairly simple set of such rules could, in many cases, do a better job of matching particular travellers' preferences than all but the most conscientious and knowledgeable human travel agents.

Natural language processing and information retrieval techniques. It is easy to imagine even more sophisticated facilities in this framework that could use whatever natural language understanding capabilities are available to parse unstructured documents into the templates used by Lens. The fields extracted in this way could then be used for automatic filtering or other processing after which human readers could look at the full text of selected articles to do more accurate processing themselves. This approach appears particularly promising for natural language documents about highly restricted domains (e.g., letters of credit—see Dhar & Ranganathan, 1986).

We have already seen how the Lens facilities can be used without any automatic natural language understanding capabilities. However, as natural language parsers become more powerful and accurate, rules like those specified in Lens will become more useful for processing a much wider range of documents.

Examples of Uses

Now that we have described the Information Lens and several applications implemented within it, we will return to the possible uses mentioned at the beginning of the chapter.

Distributing engineering change notices. It is easy to imagine semi-structured templates for different kinds of engineering change notices. The templates might include information such as: part changed, type of change (e.g., size, shape, material, vendor, weight, cost, etc.), subsystems affected by change (e.g., electrical, mechanical, etc.), description of problem being fixed, severity of problem, and explanation of change. With this kind of information structured in templates, different users could specify their own rules for selecting change notices they wanted to see and for sorting and prioritizing the change notices they receive. The change notices could still be addressed to people based on official organizational channels (e.g., "everyone in the power supply engineering group") and on the basis of informal personal networks (e.g., "I know Joe is interested in this because we talked about it in the hall last week"). In addition, however, change notices can also be addressed to "Anyone" and thus reach other people who are interested in the change but who wouldn't otherwise have received it. Notice that this approach has the desirable property that part of the responsibility for routing information falls on the receivers of the information (those who will actually use it and are therefore motivated to have it routed correctly) rather than on the senders (who are distributing the information primarily for the benefit of others).

In addition to the routing of change notices, systems like Lens could also help manage the approval process. Managers who must approve changes could have suggested response options to approve, disapprove, or ask for more information. Rules could be used to automatically sort change requests into categories based on type of request, whether it was approved or not, and so forth.

Handling software bug reports. We saw earlier how Lens could be used to help manage the process of assigning software bug reports to developers. With additional fields in bug reports such as "severity of failure" and "subsystem in which failure occured", Lens could also help developers prioritize the bugs that have been assigned to them and to select messages on other bugs they would like to know about. Also as we saw above, when a bug is fixed, the patch can be distributed to other people in the development group using a "bug fix announcement" message that has a suggested response option to load the file that contains the patch.

Distributing news and rumors to traders in financial institutions. We have seen how retrieval from external information sources, such as news wires, can be integrated into our framework. This approach could give traders access to news stories selected according to their interests, and this information would be integrated with the rest of the traders' electronic communications. At present, however, news wires provide only a limited amount of structuring (e.g., title of article, author of article, general subject area, and perhaps specific topic keywords). At least two other possibilities might make these services even more useful in the future.

First, improved natural language understanding capabilities might allow computers to parse the text of the news articles into templates (e.g., DeJong, 1979) that would allow traders to specify more useful rules. For instance, there might be templates for news events such as (1) takeover bids (with fields for the takeover target, the acquiring company, the amount, etc.), (2) oil price change (with fields for cause, old price, and new price), and (3) political upheaval (with fields for country, parties involved, etc.). This approach could lead to much more intelligent filtering of stories, but the technical problems involved in doing such natural language understanding are still quite difficult.

The second possibility is less obvious, but quite feasible technically: Human editors could append content-based templates to the news stories, just as they now append topic keywords. In this case, one human editor appends a template once for a news story that can then be processed automatically by rules for thousands of receivers. The cost of this would presumably be much less than having highly fallible natural language understanding programs operating on the same story in each of the thousands of receivers' workstations.

A final possibility for improving traders' information access does not rely on news stories at all, but instead on creating more efficient channels for diffusing the informal gossip, rumors, and other "word-of-mouth" communications that are already pervasive in financial markets. For instance, traders might use a system like Lens to inform other traders in their own firm about rumors they hear about impending price changes, political upheavals, and so forth. Instead of flooding each other with mostly irrelevant information (or relying on yelling across the trading floor), different traders could specify rules about the kinds of "rumors" they wanted to see.

Sharing "lore" about repair problems in field service groups. One simple possibility for sharing field service "lore" is to have computer conferencing systems, like those described earlier, that let repair personnel at widely distributed sites set up conference topics to discuss different kinds of problems and their experiences in solving them. Not only could people "subscribe" to the conference topics in which they were most interested, but they could specify rules to, for example, highlight discussion contributions from people who often make useful remarks.

A more interesting possibility involves using the capabilities of a system like Lens to create "inquiry networks." For instance, people thoughout the organization with specific questions could send these "requests for information" to "Anyone." Also, different people could set up "Anyone" rules to select "requests for information" and other messages about topics in which they were interested. These people might or might not themselves be the most knowledgeable experts on the topics, but they could quickly become expert at referring questions in their general topic areas to the best experts or others who could answer the questions.

One of the key ideas here is that Lens-like features allow people with questions to broadcast them to a potentially large and unknown set of people, but without running the risk of inundating the receivers with unwanted information. Receivers will only get these questions when they have specifically indicated a willingness to see questions on this topic. By providing tools to create such highly differentiated "expertise nets," Lens-like systems have the potential to greatly improve the speed and effectiveness of "organizational memories."

POTENTIAL PROBLEMS WITH SYSTEMS OF THIS TYPE

Almost any powerful technology that has the potential to benefit people also has potentials for misuse or unintended negative consequences. The system we have described is intended to help avoid some potential negative consequences of computer-mediated communication systems (e.g., information overload for individuals) and at the same time to take advantage of some even greater potential benefits (e.g., selective sharing of much more information in organizations as a whole). In order to use a new technology wisely, it is important to try to anticipate and encourage beneficial uses and to anticipate and avoid possible negative consequences. Because much of this chapter has been devoted to describing potential benefits from systems of this type, in this section, we briefly describe a few potentials for misuse and some possible remedies.

Excessive filtering. Some people, on hearing descriptions of this system, worry that it might be used to decrease the flow of information in an organization. For instance, people might use it to filter out messages personally addressed to them and thus become less responsive to information from other people in their organization. While this is, in fact, a possible use of the capabilities we have described, we believe it is an unlikely one. The system leaves completely up to each user the decisions about how cautious or how reckless to be in specifying rules for automatic deletion of the messages they receive. There are already many social forces at work in organizations that affect how responsive people are to each other's communications, and in many cases, these forces would strongly discourage people from automatically deleting messages addressed to them personally. A much more likely scenario, we believe, is that people will use the capabilities of the system to sort and prioritize messages addressed to them personally and will use automatic deletion primarily for non-personal messages addressed to large numbers of people via distribution lists, conference topics, or bulletin boards.

In this case, of course, the ability of receivers to filter out "public" messages that are unlikely to be interesting to them increases the usefulness of the public communication channel in two ways: (1) receivers are more likely to attend to communication channels whose "richness" (i.e., probability of being interesting) is greater, and (2) senders are likely to send out more information if they are not worried about incurring the displeasure of many uninterested receivers whose mailboxes would be cluttered.

Imperfect finding. Another concern occasionally expressed about systems like this is that people may have difficulty knowing what they want and don't want to see until they have seen it. How, for instance, can you find out that another group in your organization is doing something of great interest to you if they use keywords for describing it that are unfamiliar to you? Here, of course, the relevant comparison should be, not an omniscient and perfect system, but the plausible alternatives that are available. No system, including the one described here, can do a perfect job of finding all and only the information in which a given user is interested. We believe, however, that capabilities like those we have described increase the likelihood that people will find useful information they would not otherwise have encountered.

One simple mechanism for helping people find messages they don't know they want is to give them the option of seeing some number of randomly chosen messages each day. (These messages should, of course, be chosen from the "public" messages addressed to "Anyone" not from private messages between individuals.) Some of the random messages may, in fact, be of interest and may lead their recipients to establish filters that select other similar messages in the future. A slightly more sophisticated version of this approach is to have each user's rules assign a "probable interest value" to all messages. Techniques used for document ranking (such as term weighting) could be helpful for this purpose (Noreault & McGill, 1977; Salton & McGill, 1983). The system could then show a user all the messages above some "interest threshold" and a sample of other messages that are below that threshold but are randomly selected in a way that favors messages of higher probable interest.

Excessive processing loads. In the prototype version of the Lens system, there is only one "Anyone" server for all the users of the system. Clearly, when systems like this are used on a larger scale, such a single server could easily become overloaded. It is a straightforward matter, however, to have multiple "Anyone" servers spread throughout an organization, each one, for example, serving a different group, department, or division. Each of these servers can, in turn, have rules that determine when to forward messages they receive on to other "Anyone" servers elsewhere in the organization.

Privacy concerns. Many important issues of privacy and security are raised by any computer-mediated communication system that carries personally or organizationally sensitive information. These issues are, of course, important in systems like the one we have described, but they are not unique here. For instance, it is already common in electronic mail systems to restrict the audience for certain messages by addressing the messages only to specific individuals or to distribution lists whose membership is restricted. The Information Lens system uses the underlying mail system in this way and adds one more level of "public" information (i.e., messages addressed to "Anyone"). We have also implemented a simple extension to the system that allows messages to be addressed to "Anyone-in-<distribution list name>." A message addressed in this way can be received only by people in a specific distribution list whose rules select it.

There are also some intriguing new possibilities raised by intelligent information sharing systems that are not present in all computer-mediated communication systems. For example, the rules about how people filter, select, and prioritize their messages represent a new kind of potentially sensitive information that is stored in the system. Would employees, for instance, want their supervisors to know that they had filters selecting notices about job opportunities in other parts of the company? It is not clear, however, that people's rules should always be kept completely confidential. Sometimes, for instance, people may want others to know that they are interested in certain topics to encourage the formation of interest groups. There may also be times when it is desirable to tell the senders of messages addressed to "Anyone" how many people's rules actually selected the messages, without revealing the names of the recipients. Similarly, there may be times when it is desirable to display the numbers (but not the names) of people interested in different topics. Devices like this could thus provide a new kind of nonintrusive and (in some cases) nonobjectionable method for conducting instant"opinion surveys" or "market research." Clearly, careful thought is needed about when and how these possibilities are desirable, but we think the possibilites are quite intriguing.

Conflicts of interest. Most of the capabilities for information sharing that are included in the current Information Lens system can be expected to work best in communities where people share goals and where there are not strong conflicts of interest about whether certain kinds of information are worthy of attention. When there are such conflicts, for example, when an "advertiser" wants you to pay attention to something that you will in fact regard as "junk mail," then filtering capabilities like those we have described can sometimes be defeated. For instance, someone who wants many people to read a particular message can indicate that the message is about a topic that is widely interesting when, in fact, the message is not about that topic at all. It is, of course, possible to evolve filters to combat such maneuvers (e.g., "delete all messages from X, regardless of the topic indicated"), but this kind of "game" can continue to escalate with each side adopting more and more subtle techniques to filter out (or filter in) the messages. We believe that situations involving conflicts of interest like this are probably better handled by the social and economic approaches to information filtering we have discussed elsewhere (Malone et al., 1987).

RELATED WORK

There are several other previous approaches to structuring information sharing in electronic communities that have been used much less widely than distribution lists, conference topics, and keyword retrieval methods. These include: (1) using associative links between textual items to represent relationships such as references to earlier (or later) documents on similar topics, replies to previous messages, or examples of general concepts (e.g., Engelbart & English, 1968; Halasz, Moran, & Trigg, 1987); and (2) representing and using detailed knowledge about specific tasks such as calendar management or project management (e.g., Sathi, Fox, & Greenberg, 1985; Sluizer & Cashman, 1985). Our system is, in some sense, at an intermediate level between these two approaches. It includes more knowledge about specific domains than simple associative links, but it can be used for communicating about any domain, even those for which it has no specific knowledge. A few systems (e.g., McCune, Tong, Dean, & Shapiro, 1985) have used AI techniques such as production rules to reason about the contents of messages based on the presence or absence of keywords in unstructured text.

We have not focused here on facilitating the kind of real-time information sharing that occurs in face-to-face meetings (e.g., Sarin & Greif, 1985; Stefik et al., 1987) or teleconferencing (e.g., Johansen, 1984). We believe, however, that the aids we described could be useful in some real-time meetings (especially those involving very many people), and that these aids could eliminate the need for some meetings altogether.

CONCLUSION

In this chapter, we have seen how a combination of ideas from artificial intelligence and user interface design can provide the basis for powerful computer-based communication and coordination systems. For instance the use of semi-structured messages can simplify designing systems that (1) help people formulate information they wish to communicate, (2) automatically select, classify, and prioritize information people receive, (3) automatically respond to certain kinds of information, and (4) suggest actions people may wish to take on receiving certain other kinds of information. We believe that systems like this illustrate important—and not yet widely recognized—possibilities for collaboration between people and their machines. The power of this approach appears to be due partly to the fact that it does not emphasize building intelligent, autonomous computers, but instead focuses on using computers to gradually support more and more of the knowledge and processing involved when humans work together.

ACKNOWLEDGMENTS

Much of this chapter appeared previously in Malone, Grant, Turbak, Brobst, and Cohen (1987), and Malone, Grant, Lai, Rao, and Rosenblitt (1987). This research was supported by Xerox Corporation; Wang Laboratories, Inc.; Citibank, N.A.; Bankers Trust Co.; General Motors/Electronic Data Systems; the Management in the 1990's Program at the Sloan School of Management, MIT; and the Center for Information Systems Research, MIT.

5

FLEXIBLE INTERACTIVE TECHNOLOGIES FOR MULTI-PERSON TASKS: CURRENT PROBLEMS AND FUTURE PROSPECTS

Tora K. Bikson
J. D. Eveland
Barbara A. Gutek
The RAND Corporation
Santa Monica, CA

INTRODUCTION

Whether organizations are moving to network their personal computers, decentralize their mainframe environments, or build group-level computing structures, they share at least one major concern: to provide flexible interactive technology to support and augment multi-person work. This chapter reviews cross-sectional, case study and pilot research carried out by RAND's Institute for Research on Interactive Systems, which explores the deployment of current information technology in diverse user groups.

The successful integration of new technology into information-intensive work demonstrates the socio-technical properties of work groups; that is, group members are interdependent not only on one another but also on the technology, and technical and organizational issues are closely interrelated. The more advanced the information-handling tools provided to the group, the more critical it becomes to give equivalent and concurrent attention to the social processes through which these tools are deployed, and to seek a mutual adaptation rather than maximization of either the social or technical system in isolation.

Experiences in the organizations we studied indicate that even today's technologies can make multi-person information tasks more manageable, increase throughput, and permit more broadly-based and flexible work groups. However the same research suggests that realization of these benefits heavily depends on the resolution of social questions about collaboration—questions about group norms and values, equitable role structuring, and shared task management—that organizations introducing new technology are not usually prepared to address.

Advances in hardware, software, and communications expand the opportunities for collaboration. The rate of evolution in the technologies for collaborative work (for instance, hypertext-based systems for data management or joint authoring, communication systems for coordination of interactions or intelligent message handling, etc.) has probably outpaced our understanding of how such tools can be productively managed and used in organizational contexts. Taking advantage of such technologies will require some very new answers to some very old social questions.

UNDERSTANDING THE TERMS

Although the notion of work group collaboration is a familiar one, it is often presupposed rather than defined. For purposes of the research reviewed here, we found it helpful to rely on the generic concept of a "work unit" from traditional organizational research. Trist (1981) defines primary work units in the following way:

> These are the systems that carry out the set of activities involved in an identifiable and bounded subsystem of a whole organization, such as a line department.... They have a recognized purpose, which unifies the people and activities. (p. 10)

If this characterization is amended so that the work unit's activities are information intensive, it yields a reasonable starting definition of white-collar work groups, or collaborating groups of information workers.

We operationalized this definition to emphasize both the complexity and the organization of work units. That is, following Rousseau (1983), we targeted for study groups of four or more persons, representing at least two different status or occupation levels, whose activity is related by outputs or by work processes (Bikson & Eveland, 1986; Bikson & Gutek, 1983; Bikson, Gutek, & Mankin, 1987; Gutek, Bikson & Mankin, 1987; Talbert, Bikson, & Shapiro, 1983).

Work groups comprise multiple individuals acting as a bounded whole in order to get something done (cf. Dunham Johnson, McGonagill, Olson, & Weaver, 1986; Kraut, Galegher & Egido, 1986). So construed, they are inherently collaborative. This view concurs with Blomberg's (1986) in underscoring the cooperative aspect of most work activities. A group's work goal, in turn, will likely involve a number of multi-person tasks and task cycles; its activities are expected to persist over time and to survive membership changes. Finally, we emphasize missions; that is, what groups do, in accounting for cooperation. In the phrase "work group," 'work' and 'group' get equal stress (Akin & Hopelain, 1986).

From this standpoint, the goal of technological support for work group collaboration is to enhance its mission performance. This interpretation accords with accepted definitions of tools as means for extending the capability of individuals, work groups, or organizations (Tornatzky 1983). For information work, the tools are flexible, computer-based information and communication technologies that aid the completion of multi-person tasks.

More specifically, the research summarized here focuses on interactive systems that can support multiple functions and are appropriate for use by all work group members. (This is not to claim that every function of the system is appropriate for all members, but only that some subset is appropriate for each of them.) This conceptualization of work group technology remains quite broad and is satisfied by widely varied configurations of hardware, software, and communications media. Candidate systems might range from personal computers communicating via the manual transfer of floppy disks to supermicros on broadband networks.

The systems we observed fall somewhere between these extremes, although they tend to be "lagging-edge" technologies rather than the leading-edge variety. Nonetheless we suspect there is much that is generic about group work and methods for augmenting it. If so, examining experiences with today's tools can prove helpful in understanding problems and prospects for supporting group collaboration with advanced technologies.

RESEARCH BACKGROUND

Since 1982, we at RAND have undertaken a number of studies of interactive systems in organizational settings. All rely on the common definition of the technology and the work group outlined above. However, they intentionally incorporate diverse research methods. Projects on which this discussion draws most heavily include:

- two extensive reviews of the literature (Bikson & Eveland, 1986; Bikson, Gutek, & Mankin, 1981)
- a large-scale cross-sectional study of 55 work groups in private sector organizations (Bikson, 1987; Bikson & Gutek, 1983; Bikson, Gutek, & Mankin, 1987)
- case studies of multiple work groups in single organizations (Bikson, Stasz, & Mankin, 1985; Stasz, Bikson, & Shapiro, 1986);

• two field intervention projects: a pilot electronic mail project to design, implement and track a message-handling system (Eveland & Bikson, 1987) and a long-term experiment comparing electronic and conventional interaction media for support of two otherwise identical task groups (Eveland & Bikson, 1988).

Although each research effort addresses project-specific hypotheses, they share some guiding propositions. For example, the projects all assume that the work group is the critical unit of analysis. They look secondarily at the overall organizational context in which the groups are embedded, at how targeted groups interact with other groups, and at individual differences. Almost never do they examine entire occupational strata (e.g., all managers, all professionals) because these are "groups" only in a statistical sense and do not reflect the organization of work.

The projects further suppose that any interactive technologies introduced into work settings will be, following Kling and Scacchi (1982), more like webs than discrete entities. This tenet leads to a technical focus not on highly specific electronic tools but on the broader interactive environment of which the tools are a part. That environment, we believe, should be modeled generically as an information-communication system. For example, its major components can be regarded as "messages" (chunks of content, which may be composed from text, numbers, images, graphics, and so on, and which may be operated on with content-appropriate electronic tools); "senders" (who compose and transmit the messages); and "receivers" (either another individual(s) or the same individual at another time). If the basic mission of white collar work is generation, transformation, or transmission of information, then this model of the technology web would seem to suit it fairly well (Talbert, Bikson & Shapiro, 1984).

What happens when a web of interactive technology is integrated into information work? The result, we believe, is a sociotechnical system in the traditional sense: Work groups become "directly dependent on their material means and resources for their output" (Trist, 1981; cf. Bikson & Eveland, 1986; Johnson & Rice, 1987; Pava, 1985; Taylor, 1987). That is, individuals become interdependent not only on one another but also on the technology for accomplishing their mission. Although the avenues for collaborative work and the means for managing it are multiplied, new sources of variance are also introduced by the technology that pre-existing social structures are usually ill-prepared to handle.

Finally, we expected to observe the mutual adaptation of social and technical systems. That is, flexible interactive systems are modified and extended to fit the user context even as the work group is changing to take advantage of the technology (Bikson & Eveland, 1986). However, there is no straightforward way to measure the success of this process. For research purposes, we regarded technologies as well-incorporated into work groups on the basis of how widely they were used, how satisfied the users were, and how they affected the performance of group missions (Bikson, 1987; Bikson, Gutek, & Mankin, 1987).

The following review of findings from this program of research that bear on the question of technological support for work group collaboration. In general, the discussion relies on the large cross-sectional study cited earlier, complementing it with information gathered in the case studies and field experiments.[1]

WORK GROUPS

First, we learned that while the work group is a productive unit of analysis, groups differ significantly from one another. Similarly, group work should not be treated as a unitary phenomenon since what holds true of some types of groups does not apply to others.

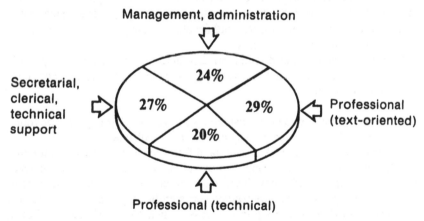

Figure 5.1.

---

1   Supported by a grant from the National Science Foundation, the study explored how well conceptions of technological innovation from previous research could inform and explain successful implementation of computer-based procedures in diverse white-collar settings. Over 500 white-collar employees, representing 55 work groups in 26 different manufacturing and service organizations, participated in the project. Data were obtained from employee surveys, managerial interviews, archival records, and observation. The research is reported in detail in (Bikson, Gutek, & Mankin, 1987) and summarized in (Bikson, 1987). For convenience, this research is often cited as the "cross-sectional" study throughout the chapter.

Given our definition of work group, it seemed most appropriate to classify groups according to their mission within the broader organization. As Figure 5.1 illustrates, our cross-sectional study of 55 white collar work groups generated four distinct functional types, each type accounting for 20 to 29 percent of the employee sample (Bikson, Gutek, & Mankin, 1987; Gutek, Sasse, & Bikson, 1986).

Management/administration: Groups in this category have decision-. making, planning, policy-setting, and oversight responsibilities. Examples in our research include corporate strategic planning offices, fiscal controllers' offices, personnel departments.

Professional (text-oriented): We distinguished two types of groups that carry out professional functions. The "text-oriented" groups were so designated because their products tend to be conveyed with textual information. Legal offices, public relations offices, marketing departments, and the like, fall into this category.

Professional (technical): In contrast, these groups tend to produce specifications, designs, formulas, models. In our study, this category included electronic design departments, internal research and development departments, manufacturing quality assurance departments, etc.

Secretarial, clerical, and technical support: Groups of this type provide support services. Examples are reservations and bookings offices, inventory control, and payroll offices.

While we initially based these categories on what work groups do, we found the four types to be associated with a number of other differences.

For example, we observed substantial differences in size and internal structure. Although the average group size overall was 10, upper management/administration groups tended to have fewer members and support groups, more members. Interestingly, both these group types were significantly more centralized than either type of professional group in the cross-sectional study (Bikson & Gutek, 1983), a finding that reappeared in network analyses of communication data in our study of electronic mail patterns (Eveland & Bikson, 1987). In contrast to previous hypotheses about size and centralization (e.g., Crowston, Malone, & Lin, 1986), these data suggest that internal structure is more influenced by group type than by size. However, within professional groups we were able to show that the smaller the membership, the more centralized its communications (Eveland & Bikson, 1987).

Perhaps more important, we learned there are characteristic sets of information handling activities distinctive of each type of group. Not surprisingly, text-oriented professional groups do a great deal of writing, editing, and rewriting; their technical peers, by contrast, take the lead in computation and database maintenance. Upper level management and administration groups develop forms and distribute information, while support groups fill in forms and process records. The activity sets delineated in Figure 5.2 were obtained by factor analysis of task check lists and provide empirical support for the initial mission-based classification of work groups (Bikson, 1987; Bikson & Gutek, 1983).

On the other hand, the same checklists revealed many commonalities across the sample regardless of group membership. As Table 1 shows, while writing original material is most prevalent in text-oriented professional groups, two-thirds of the employees in our cross-sectional sample (N=531) write from time to time as a regular part of their job. Similarly, while top management spends a higher proportion of time in verbal communication than other groups do, almost everyone reports verbal communication to be a non-negligible part of their work. And over half of all employees have some sort of information files to maintain.

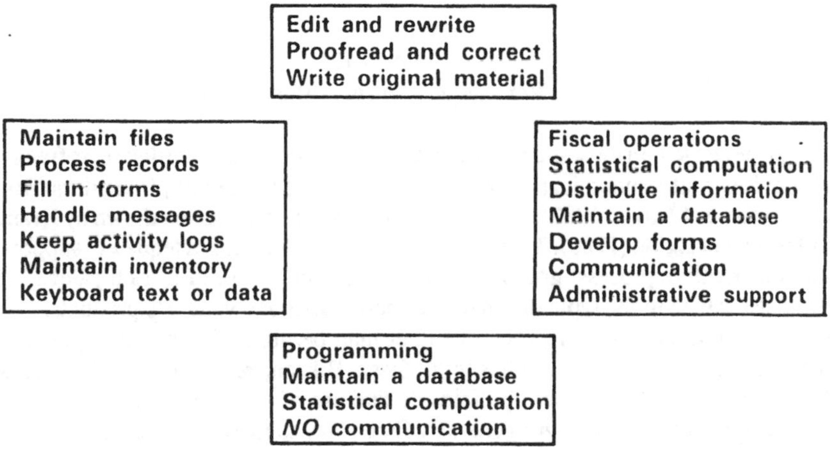

```
                    ┌──────────────────────────┐
                    │ Edit and rewrite         │
                    │ Proofread and correct    │
                    │ Write original material  │
                    └──────────────────────────┘

┌───────────────────────┐         ┌──────────────────────────────┐
│ Maintain files        │         │ Fiscal operations      .     │
│ Process records       │         │ Statistical computation      │
│ Fill in forms         │         │ Distribute information       │
│ Handle messages       │         │ Maintain a database          │
│ Keep activity logs    │         │ Develop forms                │
│ Maintain inventory    │         │ Communication                │
│ Keyboard text or data │         │ Administrative support       │
└───────────────────────┘         └──────────────────────────────┘

                    ┌──────────────────────────┐
                    │ Programming              │
                    │ Maintain a database      │
                    │ Statistical computation  │
                    │ NO communication         │
                    └──────────────────────────┘
```

Figure 5.2. Information work that distinguished groups

From this empirical look at work groups and the activities their missions subsume, it seemed our view of the supporting technology might be an apt one: a highly generic information-communication environment in which more specialized tools are embedded as needed to carry out particular group tasks.

TECHNOLOGIES

When we examined interactive systems supporting group work, we found technologies, with an emphasis on the plural. The cross-sectional study established considerable variety in electronic tools in use; even within work units, "the tech

Most common tasks	Percent who do each
● Communicate verbally	96%
● Write original material	66%
● Proofread and correct	63%
● Edit and rewrite	57%
● Maintain files	57%
● Handle messages	49%
● Fill in forms	48%
● Distribute information	47%

TABLE 1
Very general information work

nology" tends to be a loose-bundled and changing collection of hardware, software, I/O devices, and communications capabilities supplied from multiple vendors (Bikson, Gutek, & Mankin, 1987). Our data corroborate the conclusion drawn by Kraut, Galegher and Egido (1986): There is no single technology that adequately supports the collaborative process; groups rather need and make use of a "rich palette" of computer-based tools, typically involving more than one vendor's products. We add that often they do so in spite of rather than because of technology planning processes. In fact, our case studies (e.g., Stasz, Bikson, & Shapiro, 1985) suggest that even when organizational policies dictate use of a single vendor or uniform product line, work groups will generally find a way to incorporate diversity.

Hardware

To search for patterns within this diversity of equipment, we did a principal components analysis of hardware characteristics in the cross-sectional study (Gutek, Sasse, & Bikson, 1986). For this purpose, we relied on 10 archival variables: date of acquisition of current configuration, type of processing unit, availability of local communications, number of vendors involved, nature of vendor support, and who formally owned or controlled the computer system.

This principal components analysis generated four different patterns or factors that together accounted for about 75 percent of the variance in observed equipment configurations. (It should be noted that the four factors do not stand for mutually exclusive categories; rather they represent general patterns that work groups may reflect more or less closely. For instance, a work group's equipment may very closely resemble configuration 1 and also bear some resemblance to con-

2.) The four configurations are described below in order of the proportion of common hardware variation they explain.

Configuration 1: This configuration consists of micro- or mini-based systems that are owned by the organization but may be controlled by a department other than the user group's department. The system is heavily dependent on vendor support.

Configuration 2: The second configuration is typified by mini-based systems and local communications. Many vendors are involved, and equipment is likely to be rented or leased. However the system is not likely to depend heavily on vendor support.

Configuration 3: A third configuration comprises microcomputers from multiple vendors controlled by the user group; this group also has primary responsibility for their support.

Configuration 4: The last configuration, least common in our sample, is characterized by older mainframe-based systems, usually acquired before 1981. These systems are owned and supported by the organization.

We found, not surprisingly, that the four equipment patterns are not evenly distributed among work group types (Gutek, Sasse, & Bikson, 1986). For example, while vendor-supported microcomputer or minicomputer systems (Configuration 1) can be found in all work group types, they most strongly characterize the text-oriented professional groups in our research. On the other hand, technical professional groups have greatest access to multi-vendor minicomputer systems that they themselves support (Configuration 2). While this configuration is also found with some frequency among nontechnical professional groups, it is associated with significantly higher job satisfaction in the technical groups (Gutek, Sasse, & Bikson, 1986).

Support groups frequently fall heir to the oldest systems, often mainframe systems that were not initially intended to support interactive use (Configuration 4). These groups also have the most uniform equipment. Upper management and administration groups, in contrast to other group types in our sample, are not uniquely associated with any particular equipment configuration. Among all arrangements, micro-based systems (configuration 3) are most evenly represented across group types.

Finally, we examined a wide range of outcome measures in relation to equipment characteristics to determine whether any hardware properties are significantly associated with successful work group support, either generally or within work group type. We found only weak and unsystematic relationships between particular hardware characteristics (whether taken separately or bundled into configurations) and outcome measures. The one exception is access. Not having a workstation of one's own is strongly and negatively associated with system use, user satisfaction, and work group performance across group types and equipment configurations (Bikson, 1987).

Software

Because different tasks can be performed on the same hardware and the same tasks on different hardware, how well work groups are supported by their technologies is likely to be more a function of software properties than equipment characteristics. On the other hand, software arrangements are even more diverse and difficult to characterize. In the cross-sectional study, for instance, less than 20 percent of the work groups used only unmodified off-the-shelf packages. The others had modified their software environment to varying degrees, and a majority of groups (74 percent) made use of one or more applications programs written specifically to meet their needs (e.g., capital asset tracking, avionics simulation).

Faced with such an array, we sought higher order characterizations of the software in use on the basis of our conceptual framework. Given the emphasis on mission performance by work groups, we looked first to the functions that electronic tools were being used to support. These data were obtained during initial site selection. In subsequent site visits we collected more information about the software in use, asking group members to evaluate it along a number of dimensions (Bikson, Gutek, & Mankin, 1987).

We learned that functional diversity of software applications by itself can discriminate work group types. The number of basic computer-assisted tasks reported per group ranged from 1 to 8. Support groups dominated the low end of this dimension, performing significantly fewer different functions than other types of groups. Management/administration groups, by contrast, have computer support for the largest number of different tasks. In this respect, then, technology webs come to resemble task arrays; that is, quite independently of computer technology, job variety discriminates work group types in the same way (Bikson & Gutek, 1983).

Whether users performed many or few computer-assisted tasks, we found some generic software properties to be systematically associated with positive work group outcomes. Most significant among them are the following (Bikson, 1987):

Functionality, or the extent to which applications software is appropriate for assisting users' particular job functions, is a strong predictor of success. This variable is a summary measure derived from user evaluations of specific system features. Upper management/administration groups judge their software to be substantially better on the functionality dimension than other groups; support groups give their systems notably negative ratings.

Interaction support, or whether users have what they need to interact effectively with their software applications, also significantly influences work group outcomes. Another summary variable, it represents not only type of dialog with the computer but also quality of the user manual. Interaction features also generate between-group differences, with technical professionals evaluating their systems most negatively on this dimension.

Customization, or extent of modification of software to conform to work group tasks, is a third generic dimension associated with positive results. In the sample we studied, customization was sometimes provided by a system integrator or by a systems department external to the work group. In other cases, work groups themselves had the capability to develop or modify applications (using high level programming languages, application generators, user-definable keys, user-determined profiles, and the like). Interestingly, we found that professional groups (whether technical or text-oriented) are most likely to be provided with options for modifying the way their systems behave. Both upper management/administration and low level support groups are typically denied this flexibility, relying on external sources for software modifications. In any case, collaborative work is better supported by software that is adapted or adaptable to the group's tasks (as the mutual-adaptation thesis, cited above, led us to expect).

If levels of functionality, interaction support and customization affect how well a set of computer-based tools assists a work group, what can be said about the tools themselves? As indicated earlier, when groups enrolled in the cross-sectional study, we got a list of major functions served by the computer for each. We also asked work group managers to rank them in importance; the results are displayed in Table 2.

Most important	Percent of groups
Very generic applications	42%
Very specific applications	40%
Communication, coordination	18%

TABLE 2
Major work group functions for computer support

First-ranked functions for computer support across group types are either highly general (e.g., writing and editing, statistics, document preparation for 42 percent of groups) or extremely domain-specific (e.g., acquisitions decision support, product requirements analysis, customer profile development in another 40 percent). Contrary to the electronic web we had envisioned, internal coordination and communication functions (e.g., electronic mail, calendaring and scheduling, project status tracking) were rarely mentioned as important applications of the technology (Bikson, Gutek, & Mankin, 1987). Nonetheless, 44 percent of the groups used their computer system for internal messaging and 57 percent, for external mail. These data suggested, alternatively, that specific tools may be easier to see than the web in which they are embedded. In any event, we used the smaller field research projects to take a closer look at computer-based communication in collaborative work.

COMMUNICATION

A frequent question is whether electronic media can overcome physical and social barriers to enable collaboration among individuals who otherwise would not be able to work together (e.g., Feldman, 1987). We have given special attention to electronic communication in two field intervention research efforts: a pilot project to provide an electronic mail system serving the needs of RAND's nontechnical employees (see also Eveland & Bikson, 1987)[2]; and a field experiment comparing the activities of non-colocated task groups with and without electronic communication capability (see also Eveland & Bikson, 1988).[3]

---

2   RANDMAIL is a message-handling system designed to be coherent with and to enhance existing organizational communication processes at RAND. For 18 months after its introduction, message header data (to, from, and cc notes plus date /time) were captured on two Unix-based minicomputer host machines. The 69,000 message headers logged represented 800 individual sender and/or receiver nodes. Nodes were linked with organizational characteristics (e.g., department, occupation, office location) to help interpret results generated by network analyses. This research, supported by an internal grant from The RAND Corporation, is reported in Eveland and Bikson, 1987). For convenience, we have referred to it throughout this chapter as the RANDMAIL study.

3   A grant from The John and Mary R. Markle Foundation is currently supporting a field experiment to examine the utility of computer-based communication for establishing or maintaining links between retirees and those nearing retirement. Retirees and employees were recruited (N=80) from a large organization and randomly assigned in equal numbers to "electronic" and "standard" task groups, each preparing a white paper on issues in the transition to retirement. Meetings, telephone calls, postage, and the like, are supported for both groups. In addition, electronic group members are provided with networked personal computers, electronic mail, and other software. Structured interview and survey data are collected from all participants at several times during the year-long task period. Findings will be available in RAND reports and other publications (T.K. Bikson, Rand Corporation, Principal Investigator).

A tentative answer to the question, based on preliminary findings, is "yes" when the barriers are physical and "no" (or at least "not necessarily") when the barriers are social. As others have noted (e.g., Rice, 1984), computer-based communication media, even the narrow-band sort we have studied, are not simple substitutes for other channels but rather provide new avenues for the technical augmentation of work group interaction. Their effectiveness for overcoming social distance, however, has yet to be established. The creation of new collaborations that span such distances, we believe, is probably a long-term phenomenon at best.

Space Constraints

As part of the RAND pilot project, we logged message headers for the first 18 months of the system's availability. These data included the time of day that a message was sent and office locations of the individuals involved. To explore questions about spatial distance, we constructed a 9-level ordinal-distance index representing how far apart the communicating persons are in terms of difficulty in making physical contact (for example, a '9' stands for interactions between the Washington and Santa Monica offices). Figure 5.3 presents the percent of messages sent as a function of the distance between sender and receiver nodes ("adjusted" figures correct for the number of people in a physical location).

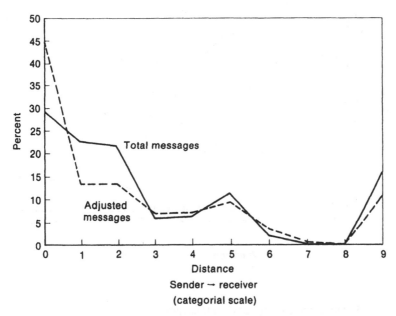

Figure 5.3. Space constraints and messaging.

As Figure 5.3 shows, we did not find that people use electronic messaging disproportionately to contact people who are spatially out of reach. On the contrary, except for interactions between the East and West coast, spatial distance was negatively associated with electronic interaction. On average, people sent about 45 percent of their messages to others in their immediate physical vicinity (Eveland & Bikson, 1987). We interpret this to mean that electronic links primarily enhance existing interaction patterns at RAND rather than creating new ones. Borrowing Orr's (1986) phrase, we seem to find "electronic hallways," but they appear in the main to parallel the spatial ones.

Time Constraints

On the other hand, the data suggest that the temporal barriers overcome by messaging are more important than is sometimes realized. In Figure 5.4, message sending is shown as a function of time of day using Pacific Standard Time. At RAND, asynchronous communication capability is, as expected, used to overcome the three-hour barrier between between the East and West coast offices.

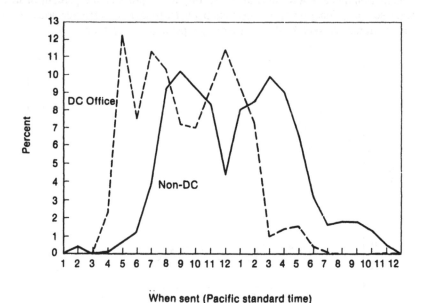

Figure 5.4. Time constraints and messaging

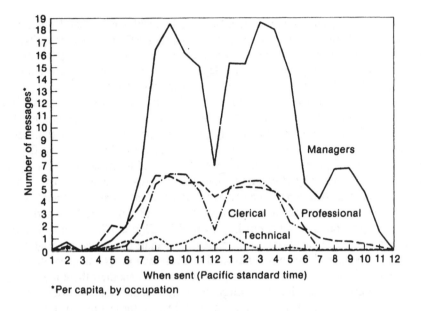

Figure 5.5. Time constraints and messaging

Even more striking, however, is the asynchronous coordination of communication displayed in Figure 5.5, where message sending times are broken down by occupational category. Different individuals within a work group appear to distribute their interactions over the work day according to occupational demands, personal preferences, and other factors. Overcoming time constraints on communication with individuals in the next office as well as in the next time zone probably deserves greater emphasis in understanding support for collaborative work (Eveland & Bikson, 1987).

Professional Groups

Organizationally, RAND is structured as a matrix with scientific responsibility partitioned between "departments" and "programs." Departments (e.g., Political Science, Information Science) are disciplinary in orientation and oversee hiring, promotion, etc.; they channel secretarial services and handle peer review for all research products. Programs (e.g., Labor and Population, Health) have a domain focus; they are charged with bringing in research funds, and they organize and oversee projects in their topic area, often involving researchers from multiple disciplines. The RANDMAIL project, for instance, drew on information scientists and behavioral scientists. Every researcher at RAND, then, participates in at least two groups. We examined patterns of communication within and between these professional groups to see how they influenced interaction and what the effect of electronic mail might be. Given our mission-oriented definition of work units, we would expect programs (which concern themselves with the production of research) to have a stronger impact than departments, in spite of what is often said about the difficulty of communicating across disciplinary lines.

A sociometric analysis of message data (Eveland & Bikson, 1987) revealed that members of different research departments, in spite of being geographically separated within RAND, regularly engage in interdisciplinary interaction. Examining the proportion of all messages sent to addressees in the same or a different department (see Figure 5.6) makes it clear that, except for messages from support staff, a sizeable majority of communications span RAND's departments. When we assessed these data longitudinally, we found evidence that department-based communication clusters became more open over the 18-month study period.

A sharply contrasting pattern is presented by Figure 5.7, which shows the proportion of all messages sent to addresses in the same or a different program of research. Sociometric analyses indicated that research programs generate communication clusters that are widely separated from one another; there is substantial intraprogram interaction, low interprogram interaction, and no evidence that electronic messaging increases the permeability of these clusters over time (Eveland & Bikson, 1987).

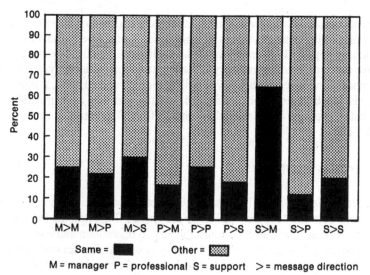

Figure 5.6. Individual message exchange: same or other department

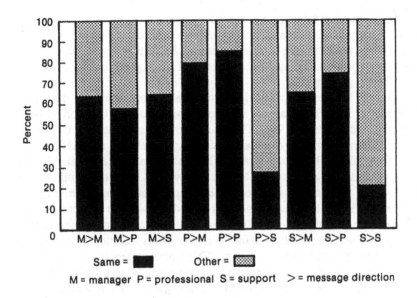

Figure 5.7. Individual message exchange: same or other program

Looking at communication among research programs by source of funding reinforces this finding. At RAND some programs (e.g., Soviet Studies, Project AIR FORCE) are chiefly supported by defense funds while others (e.g., Labor and Population, Health) receive their funds from domestic sources. When RANDMAIL interactions are examined by source of program support, it is clear that very few electronic communications span this funding boundary. This is so in spite of the fact that by disciplinary home and by type of research project, professionals engaged in Soviet Studies and Labor and Population work, for instance, may have more in common than either has with Air Force or medical research.

These analyses suggest, then, that electronic messaging is readily exploited to support ongoing collaborative work; but it does not necessarily promote the formation of new work collaborations, at least in the short term. Although electronic links can enable new collaborations as they arise for other purposes, the ability of information technology per se to stimulate new organizational interactions is still not evident.

All Groups

Like other organizations, RAND has not only professional groups but also management/administration groups (e.g., President's office, Finances, Personnel) and support groups (e.g., Telephone office, Library, Computer Services). The preceeding discussion focused on professional research groups because these collaborations permit multiple memberships and are more susceptible to change over time (and, thus, more capable of showing potential effects of new communication media). It is worth asking how, if at all, new media influence communication between different group types.

To address this question we analysed message network data twice, once using department-based professional groups and once using program-based professional groups (Eveland & Bikson, 1987). The results were virtually identical (see Figure 5.8 for a sociometric map of communicative distance between all RAND departments, distinguished by group type). These analyses indicate that, however defined, professional groups at RAND are relatively close to one another in the context of the total communication space; upper management/administration groups are also relatively close to one another; and there is very little communication between professional groups and management/administration groups. Support groups tend to be at the periphery of the communication space, not interacting with one another or with other types of groups. This pattern is a robust one unaffected by the introduction of electronic communication capabilities, a finding that may come as no surprise to students of industrial and organizational relations.

From this more focused pilot project as well as the cross-sectional study, there is evidence that electronic communication systems become embedded in the infrastructure of work and augment multi-person tasks, as sociotechnical theory suggests. Electronic mail is more a general information / communication vehicle than a substitute channel (e.g., for when the person is hard to reach physically or by telephone). Interactive linkages between work messages, work media, and workers make constraints of both time and space more manageable.

In consequence, such systems probably expand the potential for participation in multiple groups, allowing for collaborative work across a broader base of potential members. We find evidence for this conclusion in the increased interactions, within RAND, between disciplines. We also observed increased lateral interaction in our case study sites, even when it was specifically against organizational policy at the time; the organization's rules had to be altered in response (Stasz & Bikson, 1986).

A last bit of evidence comes from the field experiment now in progress (see footnote 3 and Eveland & Bikson, 1988). Two task forces, one conventionally supported and the other supported with electronic communications, were given the same general charge. Both groups of 40 members began by dividing this mission into subtasks for work by smaller groups. The conventional group spent considerable time arriving at a felicitous assignment of individuals to subgroups. In the electronic task force, in contrast, this did not arise as an issue. It was assumed that, given electronic means for overcoming the logistics of time and space for multi-person activity, members could work on as many subtasks as interested them. No one in the conventional task force affiliated with more than one subgroup, while

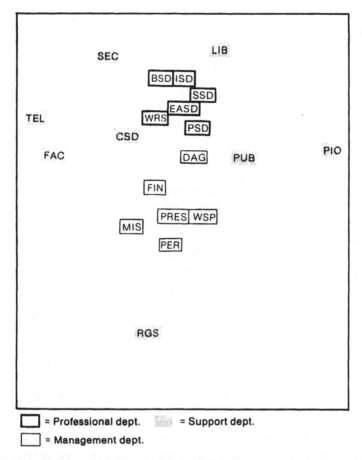

Figure 5.8. Sociogram: interactions among all groups

most participants in the electronic task force belong to two or more. At the end of the year-long work period, we will learn how these alternative arrangements fare; at this point, however, we can only conclude that availability of electronic communication media affects people's expectations as they enter collaborative work.

Interactive information-communication systems may expand the number of groups in which an individual can participate; viewed from the other side, they can expand the number of potential collaborators on which multi-person tasks may draw. But we do not see evidence that these media overcome traditional social barriers as well as they overcome space/time constraints. Rather, they seem to supplement existing preferences, opportunities, and methods for interaction.

HUMANWARE

Here humanware refers to the competence and skills to which work group members must have access in order to adapt to the collection of electronic tools that supports their collaboration. As explained earlier, as interactive systems become part of the infrastructure of white collar groups, the sociotechnical properties of their work grow in complexity and salience. We have reviewed some of the ways technologies conform to work groups. It is now appropriate to look briefly at the other side of mutual adaptation; that is, how groups change their work styles and acquire the knowledge resources to take advantage of interactive tools.

In a sociotechnical system, as we have noted, individuals become interdependent not only on one another but also on the technology. How, from the human side, is this managed? In the cross-sectional research project, we inquired of all individuals within computer-supported work groups whether or not their use of the system was voluntary. Not surprisingly, the response largely depends on the type of group. Support groups had the lowest positive response (58 percent said it was voluntary) whereas technical professionals had the highest rate of voluntary use (88 percent).

These differences, to be sure, reflect a group's organizational status and the nature of the tasks it performs as well as the desires of users. It is open to question, however, how much voluntariness will be tolerated when a work group is converting its procedures to new technologies. Interviews with work group managers produced ambivalent responses. On the one hand, managers want the transition to be voluntary; on the other hand, they make it clear that workers who are unwilling to learn to use the new tools may not be able to remain in the group.

In the RANDMAIL project we sought to develop a system useful to people who are not regular computer users. For instance, headers rely on normal names and preserve most of the regular internal memoranda conventions. We also sought to make it serve those who are not computer users at all. The database has hardcopy addresses for all employees who do not use electronic mail; when such individuals appear in a "To:" or "Cc:" line, the message is automatically printed and sent in the next internal mail distribution. Overall the new system has been a success. However, our data suggest that within the first few weeks, new electronic mail users seem to divide into frequent and infrequent users (Eveland & Bikson, 1987). Figure 5.9 displays early messaging behavior for all individuals who began using electronic mail during the 18-month logging period; that is, regardless of the calendar date at which an individual user came on line, Figure 5.9 reflects all new users' first 9 months. Over time, those in the heavier use category show gradually increasing reliance on RANDMAIL. The others (the light users in Figure 5.9) show no change, remaining minimal users throughout the study period. We do not know how the hardcopy population used mail. Anecdotal data suggest, however, that

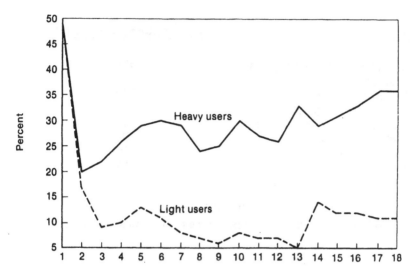

Figure 5.9. Individual differences and messages

these individuals are not easily integrated into a group that relies on electronic media for shared work.

The question of how to provide long-term learning support in work groups and secure needed technical assistance remains (Bikson & Gutek, 1984; Bikson, Stasz, & Mankin, 1985). As mentioned earlier, some work groups, notably technical professional groups, handle initial and advanced learning as well as system support largely on their own (Gutek, Sasse, & Bikson, 1986). For most groups, this way of coping with the humanware problem is not an option. They rely instead on either external centralized support or the emergence of local expertise, usually a combination of both.

Data collected in our cross-sectional study confirm the well established belief that, whenever possible, individuals prefer to consult a within-group "expert" (Bikson & Gutek, 1983, 1984; cf. Blomberg, 1986). We learned in interviews that an important advantage of local experts is that they know a great deal about the task in addition to the tool. The most often mentioned disadvantages, according to local experts themselves, are that the technical assistance they provide is not recognized and supported by management (Bikson, Stasz, & Mankin, 1985). Moreover, work groups cannot all count on having a self-selected expert.

On the other hand, most of the work groups we studied (72 percent of those in the cross-sectional sample) had access to a technical department formally authorized to provide training or technical assistance. Such departments were the first-choice knowledge resource for only about a fourth of the users (see Table 3).

SOURCE	PERCENT OF USERS
Another user who happens to be very proficient	46%
A technical expert outside the group	27%
The manager of the group	11%
Printed documentation	8%
On-line help	5%
Other	1%

TABLE 3
Where do you most frequently turn for technical assistance?

It's not the organization's problem:	36.1%
It's the organization's problem in the short run:	30.5%
There is a continuing role for the organization in this area:	33.4%

TABLE 4
What is the organization's responsibility in the area of employee adaptation to technological change?

Reasons for not drawing on this resource, explored in our case studies, include: "They tell you what to do—they don't teach you how to solve the problem"; "They don't understand the business;" "You can't talk to them." Such difficulties are not surprising in view of the fact that work groups with different task orientations generally do not communicate with one another (cf. Figure 5.8).

Better approaches to the humanware issue need to be developed if work groups are to more fully exploit the capabilities of interactive information and communication systems. Innovative examples are, in fact, available (e.g., Johnson, 1986) and others could doubtless be devised. Organizations, however, seem to be undecided about what role, if any, they should play in facilitating the adaptation of work groups to ongoing technological change. When we posed this question to managers of the work groups participating in our cross-sectional study, about a third replied that it was up to individual employees to keep their skills marketable and another third thought there might be a one-shot role for the organization bringing its employees into the computer age. Only the remaining third believed organizations would have a continuing responsibility to help develop the humanware to keep pace with advancing tools (See Table 4).

CONCLUSION

Flexible interactive technologies clearly have the capability to support and enhance collaborative work. Our research provides considerable evidence that they make multi-person information tasks more manageable and enable increases in group output as well as throughput. Most conclusions, however, are conditioned by the type of group or nature of its tasks, factors that need to be taken into account in research on collaborative work.

It is worth recalling the results of the last half century of small group research about what makes collaboration work (McGrath, 1984; McGrath & Altman, 1966):

- High skill, high ability in group members
- Good group training, considerable group experience
- Autonomy, participative decisionmaking, cooperative work conditions
- Mutual liking; group members value one another's task and social attributes, hold one another in esteem, accord themselves high status
- High level of intragroup communication

Our efforts to analyze computer-supported collaborative work do not invalidate these conclusions. Rather, by increasing the complexity of organized work and speeding up the pace of group interaction, the use of advanced information technology generates a need for creative attention to the social variables that affect multi-person tasks.

The research reported here substantiates the view that interactive information and communication media help overcome barriers of space and, even more importantly, barriers of time, especially among those engaged in common tasks. That is, these tools permit reconciling individual scheduling needs with group goals. In this way, electronic messaging can become a general mode of working rather than a substitute medium to be used in case other avenues for interaction are inconvenient.

We believe, in addition, that these technology webs permit more broadly based, reconfigurable, and overlapping collaborations. In particular, interactive message systems appear to decrease communication barriers between lateral groups in an organization and to promote shared activities. They seem to facilitate communication across disciplines and to support multiple group memberships, which typically pose as serious time-conflict problems as they do distance or telephone-tag problems. In sum, new computer-based technologies allow people to collaborate with increasing numbers of individuals like those with whom they already work.

However, they do not necessarily help overcome traditional social and organizational barriers to group interaction, such as differences in status, values, and missions. These pose formidable obstacles that are more likely to be alleviated by changes in social structure and management policy than by technological advance. Although new technology may serve to destabilize existing organizational patterns and allow for the emergence of alternative organizational designs, it does not by itself ensure the existence of such designs or mandate any particular set of social choices.

New technologies for collaborative tasks, in short, are neither simply the servants nor simply the masters of organizational design. Rather, they make it ever more critical for the organizations that use them to develop creative strategies for evaluating and balancing social and technical capabilities. Successful new modes of work group collaboration will require more social and managerial innovation than has been evidenced to date.

6

ORGANIZATIONAL ARCHITECTURE FOR DISTRIBUTED COMPUTING: THE NEXT FRONTIER IN SYSTEMS DESIGN

Calvin Pava
Harvard Business School

In 1964 Gordon Moore, a founder and now chairman of Intel Corporation, first identified a persistent trend in hardware capability. The power and complexity of microchip technology, he maintained, would double every four years. It has, and it will. Meanwhile, the cost per bit of microelectronics has declined by about 28 percent each year since 1973. Moore's Law and this plummeting cost curve pose a vexing question for those who aspire to be systems designers: What do we do with the annual bumper crop of newly available machine cycles?

No technological imperative mandates any particular answer to this question. The sorting out of system design preferences is not merely a one-way flow of ideas from technical gurus down to passive users. Like any evolutionary process, a design decision is dynamic and fraught with contention. Divergent alternatives exist, as well as competing measures to assess them. Scholars offer disparate theories; vendors market products that embody different ideas about computer designs; fickle customers grab onto a few innovations.

From this interchange of systems designers, vendors, and customers, new levels of design abstractions, or architectural domains, become ratified. And, once in place, the threshold of adequacy for new systems products rachets upward, to new levels of sophistication.

This chapter traces the process by which new architectural domains emerge. Through an extrapolation of how digital technology is poised to expand, I argue that organizational architecture presents the next domain of innovative constructs for systems design.

BOOTSTRAPPING NEW CATEGORIES OF SYSTEM DESIGN

In order to keep pace with developments in raw computing technology, collective thinking about systems design must advance continuously. By resorting to more complex and abstract constructs as guideposts for systems design ("systems should do this . . ."), designers extend the scope of potential functions that users can draw upon. This progression of ideas leads to more intricate and abstract designs for gaining and using raw computing performance.

Few ideas lead to inspirational prototypes. Fewer still attract sufficient adherents to establish a new architectural domain. Only the notions that survive beyond this stage can demarcate areas where design talent, a limited resource, might be concentrated advantageously. These areas become the "problem spaces" for innovative systems products. Soon, problem domains become mandatory considerations in systems design. Once generally acknowledged as legitimate realms of discretion, they exist as architectural domains, where critical choices must then be made.

Rather than being smooth or incremental, this process is abrupt. A number of design constructs compete for legitimation; they are screened by vendors and users, not just computer scientists, as research investments and purchase decisions rule out some potentially innovative designs and sustain others. Even the government gets in the act via DARPA and other grant programs. This loosely organized anarchy only achieves closure periodically. Through a remarkable process of consensus building, a new stratum of design considerations is established. Stakeholders in the systems industry thereby muddle through, bootstrapping a roughly shared vision of "what's next" in systems design.

A discernible lifecycle runs through this process:

1. Estrangement. There is a constant pool of adaptive ideas about improving systems design. At any one time, the occupational community of computerphiles subscribes to various ideal innovations in systems design; most perish. (Some, like neural networks, have multiple lives.) Initially new ideas emerge as wild suggestions or outrageous propositions about how to make systems better: for example, it should be possible to learn a desktop computer's software in twenty minutes; a workstation should fit into a briefcase; a file should be printed at a remote station; massive CPU cycles should be devoted to repainting a single user's spreadsheet screen; systems should be created without instruction sets, or software contain only instruction primitives. Ultimately, such ideas imply entirely new domains of choice in systems design.

2. Tinkering. Whatever its content, the new attractive and ambitious concept gains a nucleus of early adherents, whom respectable computing professionals dub fruitcakes. Nevertheless, ingenuity perseveres. Many prototypes are attempted, most in large corporate labs, university computing centers, and garages. Some prototypes work, some fail; those that function invite further dabbling.

3. Commercialization. If early versions provide demonstrated advantage, then a novel category of design problem may arise. Systems with designs that address this new architectural domain are eventually produced and offered to a market containing real customers. Another cycle of assessment and redesign ensues. Meanwhile, the pioneering design implementation drives down a ruthless cost curve, while gaining capability, thanks to Moore's Law.

4. Vindication. If the emerging products attract enough customers, other companies pursue the early entrants. The new architectural domain may finally be acknowledged and a new set of specifications for systems design and assessment results. By this time, there often exists a successful, rapidly growing new venture (initially written off as a pesky mutant) whose riches "justify" this architectural domain. Larger, established companies (which may have long resisted commercializing their own laboratory technology in this area) struggle to catch up. The trappings of professional status appear. A new category of experts arbitrates disputes about the new architectural domain; a new academic specialty, complete with doctoral hazing process and refereed journals, crops up.

5. Renewal. In time, the new architectural domain is deemed essential. Systems vendors pay homage to it, usually through meaningless statements of direction, mislabeled features, and pure vaporware. Continued innovations may still occur, but novel solutions eventually grow stale, and conventional wisdom is reexamined. The design problems triggering the creation of an architectural domain are tackled anew. Fresh approaches are undertaken, often drawing on improved technologies or positing tradeoffs that differ from the prevailing norms. In this way, a new set of heretics steps forward to revive the architectural domain's vigor.

These steps comprise a bootstrapping process, by which designers, vendors, and users hammer out new categories of systems design choices. From estrangement through renewal, the ascendance of a new architectural domain follows a winding trajectory, with little preordained. Much depends on brilliance, luck, and persistence. The entire cycle, emergence through renewal, lasts about ten years.

USER-FRIENDLINESS AS MUDDLING THROUGH

A vivid example of systems design progression lies in the architectural domain of user-friendliness:

1. Estrangement. In 1975, user-friendliness as an objective for systems design was almost unknown. Only a few developers, such as Doug Englebart at

SRI, Dick Sutherland at MIT, and researchers at Xerox PARC, confronted user interfaces as a legitimate set of design variables.

2. Tinkering. Through dedicated effort and judicious use of advancing technology, designs that maximized user-friendliness were made operative, in the form of input/output devices and new conventions of screen presentation.

3. Commercialization. After a prolonged latency, the advantages of user-friendliness proved worthwhile compared with previous system designs. Some firms, such as Lotus Development, Apple Computer, and Professional Software, prospered because of their keen attention to this new architectural domain. Nonetheless, many disputes continued to rage regarding mice (one-button versus two , mechanical versus optical), windows (tiles versus overlapped), and command entry formats (menus versus command lines).

4. Vindication. By 1983, virtually everyone had hopped on the bandwagon; all systems vendors must at least appear to embrace user-friendliness. The release of numerous integrated user-friendly interfaces with generally similar designs (e.g., Star, Lisa, VisiOn, Windows) testifies to both the peaking legitimacy of user-friendliness and the pervasiveness of mimicry. By 1987, even IBM had released a mouse-pointing device and a windowing-based system command environment.

5. Renewal. So great is today's enthrallment that we fail to consider that user-friendliness, particularly as it is embodied by the icon/mouse/windows interface, might actually become outmoded. In fact, "user- friendliness" is a moving target for a growing population of systems users, some of whom are growing less timid. Advancing technology also unfreezes capabilities which are available to support computer conviviality. Recent developments, such as voice input/output, three-dimensional liquid-crystal shutter displays, and more truly portable computers, modify the available combination of technologies that might be used to enhance user machine interaction. User- friendliness will endure as an architectural domain of systems design, but perhaps the standard approach to such designs is due for a reformation.

Ascendance of the user interface seems to have arrived just in time. It created a much needed outlet for consuming the billions of machine cycles unleashed upon the world by the advent of personal computers. The spare power of virtually any PC could be absorbed by tossing bits around to make data more presentable—not just data generated by program output, but data about machine states as well (the user-friendly operating system interface).

User-friendliness, however, was not an instance of manifest destiny. Had different preferences prevailed, all those instruction cycles could have been harnessed for other means. For instance, they could have been used to extract greater decimal precision or to support more complex batch processing applications. Alternately, they could have simply been left unexploited due to a lack of imagination. There is a constant risk of underestimating a new technology's potential. On the eve of introducing the first commercial microprocessor, Intel's marketing department projected an annual demand of fewer than 1000 units annually.

SUCCESSIVE ARCHITECTURAL DOMAINS

In its brief history, digital computing has spanned at least two architectural domains. Each spanned more than a decade, and both unfolded through a sequence of estrangement, tinkering, selling, and vindication. Their success culminated in the acknowledgment of a new architectural domain, and ultimately in its renewal by the succeeding generations of malcontents who continue to innovate.

1. Subsystem configuration. Defining the arrangement of subsystems posed the first architectural domain in computing. Subsystems architecture involves tradeoffs that allocate functions between different system elements. Although most systems have built on the foundation of basic von Neumann designs, there have been countless variations in the scale and combination of basic elements, such as working memory, compilers, register sets, and mass memory.

The new domain of subsystem architecture did not spring forth immediately. Its emergence coincided with transistor technology. Solid-state semiconductors unfroze many physical constraints that had previously limited the physical configuration of systems. A wider range of possibilities became accessible. Pioneering efforts at academic research centers, such as the Lincoln Labs at MIT, produced transistor-based computers such as Whirlwind and TX-O. These machines recalibrated expectations about the basic size and configuration of basic system resources.

Seizing the initiative, a smattering of new ventures moved to commercialize systems based on innovations in the configuration of subsystems. These firms produced a new category of products called minicomputers. Unlike mainframes, minis were often used for real-time or interactive time-sharing applications, many of which were located outside sealed-off computer rooms. A handful of these new businesses prospered, including Hewlett-Packard, Data General, and Digital Equipment Corporation, spawning a completely new tier of major computer companies.

In academia, the discipline of computer science emerged. It was distinguished from the original centers of university computation in math and electrical engineering departments by its emphasis on the architecture of subsystems configuration. Beyond exploring baseline component and algorithmic technology, where math and engineering departments already dominated, computer science set out to fathom the alternate ways that different combinations of subsystems could give rise to new architectures.

Innovation continued at the level of subsystems architecture long after minicomputers and time-sharing became old hat. Starting around 1975, there was a growing emphasis on distributed systems linked by high-speed networks. Special-

ized processors began to appear, dedicated to functions such as symbolic processing, supercomputing, and workstation graphics. In the mid-1980s, people began
to tinker seriously with multi and parallel processing and minisupercomputers. In
the 1990s, subsystems technology appears to be heading towards a renaissance in
component substrates, with new materials being fabricated into microprocessors.
As always, there exist several contending options that now seem tremendously exotic, including superconducting chips, perceptron architectures, bioprocessors, and
opto-electronic components. Thus, innovation will continue at the level of subsystems architecture.

2. **User interface design.** By the mid-1970s, another domain of systems architecture was emerging, although ten years were also needed before it gained
legitimacy. This domain entailed how to arrange input and output—the user interface—as a significant concern in systems design.

In retrospect, concern for the user interface appears to have been triggered by
new base-level technology that emerged in the 1970s—integrated circuits and
programmable microprocessors. The appearance of inexpensive, single chip computers posed a unique problem: what to do with all those new machine cycles. Unlike time-sharing systems, where access to a central processor imposed a critical
bottleneck, compact microchips dispersed processing power everywhere. But how
could it all be gainfully consumed?

By 1977, a new category of machine had been concocted—the individual
workstation or personal computer. The emergence of PCs and the growing interest in user interface design reinforced each other. Personal computers covered the
computational overhead needed to support less mysterious interfaces. Easy-to- use
computers were essential to the industry's growth. With lower volume and higher
prices, more of the potential market for computers consisted of inexperienced users,
who were unwilling to immerse themselves in the arcane dialects of computerese
required to operate earlier minicomputer and mainframe systems. As a new architectural domain, the user interface was trivialized by most established systems
vendors. Initially, the firms who prospered were new ventures involved with personal computing, such as Lotus, Apple, Microsoft, and Tandy, or new firms that
built microprocessor-based workstations for specific tasks, like Wang and its word-
processing systems.

Older, larger firms of the computer establishment caught on much later. Some,
particularly IBM, excelled (at least on the commodity hardware end), although few
such companies seemed to understand PC software. Even PC software companies
were occasionally caught flat-footed. For example, VisiCorp, which offered the
first spreadsheet program, lost out to Lotus Development Corporation despite its
early market domination. Lotus simply built a more friendly spreadsheet program,
drawing on the additional power offered by a new generation of personal computers.

From 1975 to 1985, this first epoch of user interface architectures primarily dealt with the invention of convivial workstations. This effort has come to settle on the icon/mouse/windows paradigm for interface design, which was initially formulated at Stanford Research Institute and Xerox Palo Alto Research Center. With convergence on this formula now complete, the domain of user interfaces must demonstrate its resilience by sustaining innovation. Signs exist that progress will continue. The period between 1985 and 1995 may see the first real implementation of voice input/output and animated interfaces. By the mid-1990s, interfaces with genuine artificial intelligence may become practical.

THE NEXT BEACHHEAD

The progress of architectural domains is gradual; each domain provides the eventual building blocks of its successor. Subsequent domains solve problems created by the success of prior ones. The established domains of subsystems arrangement and user interface design exemplify this pattern of succession. Each emerged by drawing upon and recombining the advancements that were stabilized in the epoch preceding them. The opportunity to rearrange subsystems grew from the initial establishment of stable building blocks that employed "von Neumann" processing designs, along with the infusion of transistor component technology. Together, these early advances raised the problem of subsystems configuration implicitly, in terms of how to package these building blocks in practical ways.

The user interface became malleable, and problematic, thanks to the prior appearance of microprocessor technology, experience with the early graphics processing on minicomputers, and the diffusion of personal computers. The problem of how to make computers easier to work was thus highlighted.

If they are truly viable, established architectural domains renew themselves, as practitioners keep pushing ahead and innovating. The domain of subsystems configuration did not end with time sharing. New possibilities, like distributed processing, were developed as subsequent innovations. These later developments complement the strides made in other architectural domains. This layering of successively more abstract domains resembles other processes of evolutionary development, whereby prior accomplishments may become the elemental components of later ones.[1]

---

[1] For stunning ideas about such evolutions, see Gatlin, (1972). Her analysis suggests that certain constraints in baseline systems technology in fact accelerate the proliferation of variety in the kinds of overall systems actually developed.

As an evolutionary change that lasts ten years in an industry that jolts itself daily, the ascendance of a new architectural domain is barely perceptible until after it has occurred. Fred Emery and Eric Trist noted this phenomenon in their classic analysis of emergent social change: "We have to live for some time with the future before we recognize it as such." (Emery & Trist, 1973, p.25) Because it is imperceptible, most debate in the field of systems design concerns innovation within the established architectural domains, and only hints at any broader architectural evolution. Only after vindication, is a new set of design considerations widely recognized.

The thrust of this process is to diffuse computing technology by expanding the scope of systems design, to deal with successively broader problems. At the heart of it all lies base-level computing component technology. A chip or single-board computer with nothing else, however, is functional but worthless. The architectural domain of subsystems configuration invokes an additional level of consideration. This extends design into choosing the relationships of systems components with each other. By introducing these additional degrees of choice, computing technology becomes more than a laboratory curiosity.

Next, the user interface domain adds yet more degrees of freedom. It required choosing how to arrange input and output for easy operation. This further extends the range of issues that systems designers must confront, while expanding the market potential.

Continued improvements in the cost and capability of base-level technology will soon foster another realm of design concerns.

THE RISE OF ORGANIZATIONAL ARCHITECTURE

The next architectural domain is now forming. Like the two that preceded it, this domain is starting imperceptibly. It builds upon capabilities that were established previously, infusing a new set of considerations that extend systems design into a context more general than before. At first, these considerations may have seemed outlandish or impractical. But initial commercial success will prove that they are essential for the continuing diffusion of computer technology. Any problem area contending to become the next architectural domain in systems design must surmount two hurdles. The first hurdle is quantitative. Solutions implied by the new domain should be computationally expensive. They should consume a high volume of machine cycles. An abundance of these cycles become available continuously thanks to ever-improving circuit technology, as codified by Moore's Law. Consumption of cycles however, is a necessary but insufficient criterion, since there exist numerous black holes down which to pour extra machine cycles.

There is a second qualitative, but more essential hurdle. The candidate for the next new domain must appropriately broaden the agenda of systems design. It should reduce a major, pending bottleneck to technology diffusion arising from the success of prior architectural domains. Specifically, the problems created by advancements in subsystems arrangements and user interface designs should yield new areas where the next wave of architectural innovation must focus. Recent developments that raise new problems include:

- **Computers become networks.** Recent advances in subsystems arrangement have fostered a radical, thorough distribution of computing resources. As processors become more specialized for particular tasks such as database processing, array processing, and symbolic computation, the integration between systems intensifies. New products in network file systems, remote procedure invocation, and distributed systems software are transforming the character of networks. No longer just a conduit, the network is becoming a fundamental extension of the internal system bus. Today, rather than shipping files, networks provide a dynamic medium for distributing computation. Also, with analog signals being digitized, communications is truly merging with computing.
- **Hi-fidelity Interfaces.** Recent user-interface designs have finally driven some of the mystery from computing, at least for conventional single-user desktop machines. The underlying base-level technologies for creating more friendly interfaces are continuing to develop. Striking progress will occur in voice recognition, natural language processing, and graphics presentation.

These advancements will make obsolete the current formula of "mouse/window/icon" interface designs, replacing them with higher fidelity interfaces. These improved interfaces will make more sophisticated and complex systems simpler and more appealing. Without new high-fidelity interfaces, it will be difficult to access far-flung network resources, or, to tap the potential of forthcoming applications, such as massive distributed data bases. New user interfaces will lower the barriers to easily using computing networks, just as they opened personal workstations to a wider audience.

In the future, computing will therefore involve very few computers, at least as we are accustomed to thinking of them in terms of concrete discrete devices. Instead, they will be reincarnated as networks, or "complex, heterogeneous, multinode computers." These intelligent networks will harness a variety of processing capabilities. This generation of equipment will be unlike the telephone switching system, a technological marvel that shields its users from complexity. The next wave of network computing will not be completely opaque. Its power and capability will be accessible, enabling people to tap complex information and to reprogram the network's computing operations according to their needs.

Network computing is not just a wild possibility of the distant future. Network computing has already started arriving incrementally, in products such as Digital's Vaxcluster, Sun's NFS, IBM's APPC support, and genuine, large-scale distributed application development products such as Ingres and Sybase. Higher-fidelity interfaces also seem imminent. New pointing devices superior to the mouse will soon be offered. Powerful hardware that can support three-dimensional graphics is also due.

With computers becoming networks and interfaces gaining fidelity, a new opportunity for systems deployment appears. A different limit arises that prevents reaping actual benefits from network computing. This limit is *the match between collective activities and the capabilities of distributed information systems.* These barriers to the successful commercialization of "friendly network computing" lie in the domain of *organizational architecture* for computing systems. This design domain concerns how collective enterprises and their distributed systems intersect at the boundary between information technology and the organization of work. In essence, *the organization of an enterprise is treated like another level of systems interface that needs to mesh with other systems elements.*

Systematically modifying organization arrangements will tap greater advantage from proposed distributed computing applications. Practically speaking, systems design must become more comprehensive by melding systems design and organization design.

The benefit of joining organization design and systems design is at least partially suggested by experience in manufacturing automation. Since 1981, a number of leading industrial firms such as Procter & Gamble, Cummins Engine, Ford, and General Motors, have been committed to significantly modifying the conventional approach to organizing their factories. The reason is that by thoroughly renovating factory organization, new technology is better assimilated. In cases like General Motors' U.S. joint venture with Toyota, modified organizational arrangements actually yielded greater performance improvement than did reliance only on advanced technology.

Undoubtedly, the domain of organizational architecture seems a bit ethereal and idealistic, perhaps like concern for the user interface sounded in 1974. But just as the personal computer eventually vindicated the user interface, network computing will validate organizational architecture. Distributed systems that match their organizational milieu will succeed. Distributed applications lacking a design that mutually integrates computer architecture with organizational architecture will fail.

Signs of organizational architecture should be appearing already in the form of new products and techniques that increase the choices that are available at the design level of organizational architecture.

THE APPEARANCE OF ORGANIZATIONAL ARCHITECTURE:
MAKE VERSUS BUY

Systems development efforts pivot on the distinction between make versus buy. Ideally, a firm could buy off-the-shelf systems products that would exactly match its needs. This, however, is rarely the case. Instead, firms purchase components, or semicustom products, from which they make solutions. Between make and buy sits the MIS function, as well as the occasional outside contractor who serves as systems integrator.

As the organizational interface grows, rudimentary products appear on both sides of the make/buy boundary. These early incarnations suggest the future importance of organizational architecture in systems design.

On the "buy" side, new products that are pertinent to organizational architecture fall into two related categories. First, there are "intersystem substrates," or systems software and hardware that provide an underlying foundation for distributed computing, such as local area networks, distributed operating systems, remote procedure protocols, and file format translators. Elements of these substrates have been delivered into market since 1982, such as Xerox Ethernet, Digital Vaxclusters, Tops networking, Sun NFS, and the MIT-originated X- Windows package.

A second category of emerging products is "groupware," which consists of distributed applications that run across networked computers. Transaction management systems are relatively passive; they execute and clear a spot exchange. However, groupware applications may cut across a patchwork of organizational boundaries, and they evoke elaborate sequences of mutual adjustment between people or computers. Early groupware products tended to focus on large, dedicated production operations, where costly integration could yield acceptably large payoffs, like materials requirements planning (MRP) systems and newspaper editing systems.

Most groupware offerings failed to capitalize on organizational architecture. Material Requirements Planning (MRP) products, for example, were designed with little regard, if any, for organizational architecture. They presumed that an effective factory already existed on the customer's site and that its organization need not change to use the additional information generated by MRP. Considerable frustration resulted, as customers discovered that technology by itself improved performance very little.

At best, a few prior groupware products took account of existing organizational architectures. The Atex newspaper editing system, for example, was carefully designed to preserve the professional protocols of big city newsrooms. Reporters had exclusive access to their private computer files, much like the private file drawers they maintained in their own desks. The emphasis was on preserving conventional newsroom decorum with no consideration of organizational changes that could gain more advantages from new technology.

Another small portion of these initial groupware products, like the Participate computer conferencing system, were not targeted to any particular functional operation. Instead, they served as a neutral, content-free medium for the amalgamation of individual efforts.

In short, no early groupware products aimed to jointly modify the organizational architecture in synergy with computing technology. The best some aimed for was to match existing organizational arrangements.

Computers today are increasingly connected via networks, attracting a new breed of groupware into the market. Compared to earlier offerings, these programs are higher volume, more general purpose products. Rather than specific functional applications, they serve as "groupware construction kits" that support collective work. Some products, such as Higgins and the Coordinator, emphasize communications and the need for follow-up or information-sharing among members of a project or work group. Others, such as ForComment, target a generic collective task like drafting documents. At a slightly higher level of generality, there exist distributed database products, like Oracle and Sybase, which provide a limited but sophisticated development environment for groupware applications.

On the "make" side of systems development, there exist new design methods that can help people shape organizational architectures. Design methods provide vital support to internal MIS groups and third-party contractors in putting together new distributed systems. A method guides design teams by providing a common agenda of questions as well as a shared terminology. Methods instill comprehensiveness into design work, by mandating consideration of different perspectives. Good methods help systems design avoid the quagmire of user participation for its own sake by imposing a beginning, middle, and end on the interactive phases of design.

One approach to jointly designing technology and organization that seems pertinent to distributed systems has been developing since 1949. It is called *sociotechnical design*. In addition to joining technology and organization into a mutually beneficial combination, socio-technical design includes specific analytic methods to scrutinize work, redesign organizational arrangements, and propose technology enhancements. Recently extended beyond factories to office settings (Pava, 1983), socio-technical design can deal with both explicit, linear production processes that convert raw materials or information in a stepwise fashion, and less obvious, nonlinear processes of deliberation, where people collectively define and resolve difficult tradeoffs.

Deliberations contrast with the more predictable steps that convert materials and information in factory-like settings. Often, these ill-defined, recurring conversations serve as the grist of work in managerial and professional organizations. They are increasingly common in automated factories where routine conversions are monitored automatically. Deliberations function as a substrate of ongoing communication and exchange. This background process sometimes precipitates decisions. Hence, when applied to office settings and to collaborative work in offices, socio-technical design can offset the conventional fixation on decisions which often drives the design of information systems.

The potential for socio-technical methods to infuse traditionally technocratic approaches to systems development is not that outlandish. Virtually all of the U.S. industrial plants mentioned previously that have been reorganized to better deal with new production technology have been modified using socio-technical ideas and methods. Lately, system vendors have shown growing interest. In 1986, the sales growth of distributed computing equipment flattened due to a slump in computer industry demand and the inability of customers to absorb new distributed systems. Several major systems vendors began investigating variations of socio-technical methods to help accelerate the sales cycle. Using this analysis, sales representatives can help customers specify their needs and then help modify their organizations accordingly.

In short, there exists a reservoir of new developments on both sides of the make/buy boundary that promise to help confront the new domain of organizational architecture. The need to do this will escalate as distributed systems become more common. Because no two enterprises are organizationally identical, it seems likely that effective organizational architectures will require a high level of customization, even when using ostensibly standard products.

IMPLICATIONS

The ascendance of organizational architecture and the related innovations in products and methods imply fundamental change in a number of related areas, for example, the disciplines taught in educational programs, the role of functional specialists, and the template of corporate organization.

Educational Programs. Acknowledgment of organizational architecture will legitimate often neglected variables as new, mandatory considerations in systems design. Specifically, organizational arrangements concerning structure, job definition, procedural flows, human resource systems, and cultural norms will be factored into the design of distributed systems. For instance, remote sales support systems built by competitors in the same industry may work quite differently because of their contrasting approaches to sales recruitment, hiring, development, supervision, and strategy.

Educational programs that train systems designers will need to expand the scope of their curricula accordingly. Topics relating to organization design should be incorporated into teaching and research, just as materials relating to perceptual psychology and visual presentation were added to support improved user interface design. Conversely, organizational behavior and human resources curricula will find it useful to add materials pertaining to distributed systems design.

Staff specialist roles. Expanding the scope of design considerations requires additional expertise. Specifically, to design organizational architectures, MIS staff will want to draw upon specialists in human resources and organizational effectiveness. In the near term, teamwork will be important to develop and implement new distributed systems. New design methods can help foster this collaboration.

For the long run, however, this pooling of expertise must take place on more than a project-by-project basis. Every firm consists of one unified information-processing architecture. Part of this architecture is computational, part of it organizational. Therefore, it would be sensible to establish a unified Corporate Information Architecture Group. This function would consist of MIS and human resource professionals who use integrative design methods and draw upon a high level of cross- training. Rather than being housed in a remote functional office, this group would be more usefully located in a specific business unit. Line management "customers" would expect them to sustain a complete working information architecture, instead of just delivering hardware and software. Individual systems development projects would be de-emphasized, with priority shifting to continuous, ongoing capability improvement programs that involve modifications to both technology and organization.

Corporate organization. Improvements in basic capabilities "trickle up" to create changes in larger scale patterns. Organizationally architected network computing will lead to a shift in the structure of corporate activities. The flexibility and capacity of small discreet units will rise, as the costs of coordinating work between them plummet. A more cellular pattern of organization becomes optimal, with small focused units and quick temporary linking to accomplish larger, overall tasks.

Inside companies, this cellular pattern takes hold in the form of disaggregation and lower utilization of central services. Already, the prevalence of desktop computing is decreasing reliance on central services such as graphics departments and typesetting shops.

A cellular format will also develop between companies, which will change the template of corporate organization. To compete in today's quick-changing markets, companies will define make/buy boundaries less rigidly, so as to capitalize on the greater speed and expertise of selected outside vendors. In manufacturing, this is evident in the growing reliance on outsourcing, with close coordination supporting "just in time" supplies. Similarly, companies are establishing more strategic partnerships that link small, focused development groups across large corporations.

Relying more on networks than hierarchies, companies will shift into a new format of multinode organization. (Beer, 1972; Miles & Snow, 1986; Pava, 1985; Piore & Sabel, 1984) A multinode enterprise is a network of firms that conditionally link together to undertake a particular endeavor. To coordinate, these firms rely on a variety of mechanisms including shared values, spot contracts, strategic alliances, development agreements, joint investments, common threats, licensing, and distributed computing systems. More than one multinode can exist in a single industry, and one firm can participate in many. Despite the collaboration required, competition would remain strong, as firms jockey to dominate the direction of a multinode's development and to maximize the profits extracted from it.

A network-like structure between corporations may at first sound exotic, but precedents do exist. Auto production and aerospace programs in the U.S. often assume this pattern, as do Japanese zaibatsu corporate families. But the multinode is distinguished by its vague distribution of responsibilities, its patchwork of linkage and control mechanisms, its extensive reciprocity of innovation, its high instability of member firms, and its multiplicity of capital sources. To a certain extent, firms have always relied on networks of relationships with other companies that were only partially formalized. But in multinode configurations, there exists a far greater reliance on unique, substantive contributions from companies besides one's own.

As organizational architecture becomes a growing element of systems design, other related factors will need to unfreeze and change. Potential alterations could take place in areas ranging from educational programs and staff specialist roles to the basic form of corporate organization.

CONCLUSION

I have argued that organizational architecture will serve as the next beachhead of innovative systems design. This emphasis will develop thanks to new challenges posed by the triumphs of the two preceding architectural domains of subsystems configuration and the user interface. Their legacy is networked computing with high-fidelity user interfaces. This new computational environment requires considering organizational architecture as an integral part of systems design. As a result, the design of systems and organizations will converge and begin to mutually constrain one another. Systems must be organizationally architected, whereas organizations, to be capable of assimilating this new technology, must be informationally architected to generate valid information and seize it.

Rather than technology alone, it will be the integrity of a singular information architecture, comprising both technology and organization, that begins to spell the difference between prosperity and failure. Innovations at the level of subsystem configurations produced time-sharing minicomputers that put one computer to work with many users. The ascent of user interface design was triggered by personal computers, which created a different ratio of one person with one computer. Next, the organizational interface will grow with the deployment of network computers, and a new unity will be established that mandates one organization with one information architecture.

7

DEVELOPING THE MANAGEMENT SYSTEMS OF THE 1990s: THE ROLE OF COLLABORATIVE WORK

Paul M. Cashman and David Stroll
Digital Equipment Corporation

INTRODUCTION

The purpose of this chapter is to draw a connection between the problems of managers trying to sustain organizational performance in increasingly complex environments, the kind of information technology (IT) support that may enable managers to achieve sustainable management of complexity, and the likely impact of such technology. We pay particular attention to the role of collaboration, both as a key aspect of the nature of managerial work, and as a driver of requirements for IT support of managers.

We begin with Huber's analysis of the nature of the post-industrial environment, in order to emphasize the sources of complexity in the business environment. We then review three recent studies of IT support for managers, which together show that at present, management work is poorly supported by computer-based systems. We trace this poor support to the lack of a holistic IT-oriented perspective on management work, and propose that this gap can be closed by recognizing the importance of managers' structure-creating and structure-manipulating activities. We then discuss the system requirements that derive from this view, leading to a description of our efforts to prototype a Management System meeting these requirements. Finally, we project the expected effects of a Management System.

THE CHALLENGE OF GROWING COMPLEXITY

Today's business environment is complex and becoming more so. Huber (1984) analyzed the post-industrial environment independently of current trends, and characterized it as marked by more and increasing levels of knowledge, complexity, and turbulence. The increase in knowledge will lead to increased specialization of components such as people, organizations, products, and services. Complexity is a function of the diversity of the components, the number of components in a system, and the number of relations between components. Each factor of complexity will increase: Diversity increases due to the increase in knowledge and hence specialization; the number of components increases due to the need to harness more specialized components to accomplish a task; and the number of relations increases due to increased interdependence, a consequence of component specialization. Turbulence will increase because events will be shorter in duration, meaning more events occur per unit time, and there is an increased likelihood of an event's affecting a particular (sub)system due to increased interdependence.

Huber considered the problem of organization design in the post-industrial environment and suggested that increases in knowledge, complexity, and turbulence lead to needs for more and faster organizational decision making, more and faster organizational innovation, more continuous and wide-ranging information acquisition, and more directed information distribution (to avoid overload). Huber proposed several possible organization design strategies including scanning and probing units for information gathering, formalized handling of nonroutine information, use of top management teams, and formal decision process management (i.e., managing decisions like projects).

Given the nature of the post-industrial environment, the real challenge to management in post-industrial organizations [1] is to achieve sustainable management of complexity (Cashman & Stroll, 1987). A business organization must sustain or increase its level of performance in the face of an increasingly complex, turbulent environment. Effective use of information technology, especially to support the work of business and senior corporate managers, will be a key component of this effort.

THE CURRENT STATE OF IT SUPPORT FOR MANAGEMENT WORK

Recent studies show that present-day IT provides little support for managers and management work.

The Executive Research Project (Stroll & Miller, 1984) surveyed 39 senior executives in ten companies, where the executives ranged from single-function managers within a business unit to managers of portfolios of businesses. Although almost all of the managers had personal computers, very few used them for anything other than checking stock prices or personal financial planning. Their general perception was that computers did nothing truly useful for them, and that their work

---

1   We use this phrase to denote organizations or suborganizations that exist in environments that
     can be described, according to the above criteria, as post-industrial.

did not lend itself to the kinds of applications currently being developed or sold. Although they felt that many executives were intimidated by the thought of using computers, they themselves were quite willing to try a prototype executive workstation, providing it met their requirements.

A 1986 study of over 900 middle managers and professionals by Jack Nilles and his associates at the Center for Futures Research at UCLA (Nilles, El Sawy, Mohrman, & Pauchant, 1986) examined the impact of a range of information technologies, including telephone, traditional internal and external communication media, voice messaging, computer conferencing, and computer technologies (text processing, personal computing, spreadsheets, data management programs, and specialized programs). They found that "(c)omputer technology is less generally supportive of managerial roles than any of the other technologies we considered, yet it is the most supportive of professional roles" (p.4). This was due to the tendency of the computer technology to support diagnosis and reflection more effectively than action. The former attributes were more characteristic of professional roles, and the latter of managerial roles, according to the study's authors.

In 1987, Lotus Development (United Kingdom) and the Institute of Directors (an association of directors of British companies) commissioned a study by Wharton Information Systems of personal computer use by directors of Times 1000 companies (Wharton, 1987). While "few directors today fail to appreciate the importance of information and its efficient use to successful operation," only 27% of directors (half of them finance directors) and 14% of CEOs used personal computers. Most of the usage had only begun within the previous two years. Intimidation by computers and the inability of technical professionals to effectively communicate with directors played a role in keeping these numbers down, in part because these factors contribute to a situation where "the requirements of directors and the applications which might be of greatest interest to them are not readily understood." The directors themselves overwhelmingly felt that they were too busy to learn to use computers, and that it was easier to ask an assistant. This suggests that the directors, like their colleagues sampled in the Executive Research Project, perceive the applications available on computers as being of little value to them in doing their work.

TOWARD A HOLISTIC IT-ORIENTED VIEW OF MANAGEMENT WORK

Current IT-Oriented Perspectives on Management Work

These studies suggest that today's computer-based tools are insufficient for the task of supporting management work. Poor communication between senior managers and technical professionals, as noted in the Wharton study, would not account so much for the sheer absence of effective tools as it would for such tools' not being deployed if they did exist. The absence of effective computer-based tools for managers stems, we believe, from IT providers' not viewing management from a systemic or holistic perspective.

There appear to be two related perspectives from which people analyze management work with a view toward providing IT support. The first perspective is that of the manager as decision maker. Managers are seen to engage in solving unstructured or poorly structured problems by making decisions. This has led to a large body of research on decision support systems (Keen & Scott Morton, 1978), group decision support systems (Kraemer & King, 1986), and expert support systems (Luconi, Malone, & Scott Morton, 1985).

The second major IT-oriented perspective on management work may be called the information processing perspective. This perspective focuses on the large number of information processing activities a manager engages in daily, the brief duration of any given one, the incessant round of meetings and phone calls, the demands for more information, etc. This perspective has led to development of executive support systems (DeLong & Rockart, 1984; Rockart & Treacy, 1981, 1982; Scott Morton, 1983) for viewing the "hard" and "soft" data used by managers, calendar programs, and attempts to effectively use asynchronous forms of communication such as electronic mail and computer conferences.

Both perspectives are rich in insights, but neither perspective alone is sufficient. In the next section, we break down "decision making" into a number of constituent activities, and show that neither perspective views decision making holistically, and consequently a system based on one perspective or the other would omit support for many activities in which managers engage. After this analysis, we will return to the question of a holistic IT-oriented perspective on management work.

An Analysis of Decision Process Management

Decision process management, an organization design element proposed by Huber (1984), is simply the application of project management concepts and techniques to decision making. Decision process management is necessary because of the criticality of decision making to an organization's success (Huber & McDaniel,

1986) and to the number of activities and components that must be engaged in making complex decisions.[2]

In this analysis, we break down the organizational decision making process into a number of constituent activities. Our underlying contention is that IT support for management in post-industrial organizations must not only support most individual activities, but must support them as a system, i.e., as an integrated, extensible, consistent set of capabilities.

The classes of activities that need to be managed in a decision process include:

- managing awareness of the need to make a decision
- empowering people to make the decision
- information gathering and dissemination
- problem structuring
- modeling the problem situation and decision options
- making the decision
- acting on the decision
- managing the decision process.

It is worth noting here that since these activities must and will go on regardless of the form of organization in which they occur, so too must IT support these activities, regardless of the form of organization. In other words, although the organizational structure may reduce the complexity with which components of the organization must deal (Fox, 1981; Galbraith, 1973; Malone & Smith, 1984), on the micro-level of personal and small group activity it will be necessary to carry out and support the above classes of activity. A corollary to this point is that IT support for these activities must be capable of being reorganized at least as quickly as the human organization can be reorganized.

Managing awareness of the need to make a decision. This activity might better be called "managing awareness of a problem or opportunity," and may originate anywhere in the organization and at any time. The awareness must be communicated to the people empowered to make a decision, or who can empower others to make it. Managing awareness may require setting up the proper organizational structure, processes, rewards, and culture to encourage people to be alert, take risks, and communicate.

2   Our analysis of the IT support requirements for managers in complex environments is not tied to the decision-making paradigm of organizations (Huber & McDaniel, 1986), or any other organization-theoretic paradigm. We choose to analyze activities involved in decision making because the constituent activities are representative of the activities that go on in organizations. Also, it is not our intention to propose or endorse specific organizational changes to improve effectiveness in complex environments.

Empowering people to make the decision. This may be done through delegation, through a policy decision (for decisions of a given type), or simply through not stopping subordinates from working on an issue, but reviewing their progress toward a decision. Empowerment is dynamic, in that policies may be measured for their effect and subsequently revised, authority once delegated may be revoked or transferred, etc. It is not always clear to some person or group seeking empowerment just where or to whom to apply.

Information gathering and dissemination. Information gathering and dissemination will be continuous in complex, turbulent environments, and may require the establishment of special units for environmental scanning (El Sawy, 1985; Ghoshal & Kim, 1986) and message passing (Huber, 1984). In addition to the information processing activities of scanning and probing the environment, the needs of the information consumers must be communicated to the information providers, the results transmitted to the consumers, and feedback on the quality must be transmitted back to the providers.

Keeping the information current is also a key activity. As the problem or issue moves toward a decision, it will be necessary to keep people informed of new developments that might impinge on the decision. Also, the information gatherers will have to be apprised of the state of the decision process, to better enable them to focus their gathering and disseminating.

Problem structuring. Research shows that different levels of managers progressively structure a problem by dividing it into subproblems, linking the subproblem solutions, and imposing constraints (Humphreys 1984; Humphreys & Berkeley, 1983; Phillips, 1983). In operational terms, the subproblems, constraints, and the like must be communicated and their solutions coordinated.

Modeling the problem and decision options. As decision makers strive to understand and structure a situation for the purpose of making a decision, they may try to model the problem, decision options, possible consequences of these options, and preferences for outcomes (Phillips, 1985). Activities under this heading include identifying needs for, and sources of, data or knowledge with which to construct the model, casting the data in the "right" form, constructing the model (which may involve coordinating a team of specialists), running the model and testing its assumptions, interpreting the model output (which again may involve specialists), and feeding back the results of the interpretation into the model construction process.

Making the decision. Making a decision is often a compromise between an "ideal" rational solution and the necessary tradeoffs due to political influence, hidden agendas, corporate culture, and power needs of individuals or groups. Also, large decisions are often the product of a stream of smaller decisions, such as how to frame the question which will be decided.

Acting on the decision. Once a decision is made, it should be acted upon. Commitments to action are made and must be monitored, including renegotiations of commitments as circumstances change (Flores & Ludlow, 1981). Actions by different groups, perhaps widely dispersed geographically, must be coordinated, possibly over a long period of time. A particularly difficult problem for managers is coordinating plans, not just actions (Stroll & Miller, 1984). In coordinating plans, it is not always clear what the dependencies are; moreover, a turbulent environment mitigates against stability of plan assumptions, and contingencies cannot always be enumerated.

Managing the decision process. Managing the decision process involves coordinating the work of numerous specialists, monitoring the environment and the participants for changes that may affect the assumptions on which the decision and subsequent actions are based, communicating the partial results of the process to the appropriate people for information as well as coordination,[3] and keeping the process "live." Liveness means that the process is understood, accepted, and adhered to by the participants, and also that it is firm enough to provide useful guidance, but flexible enough to accommodate changing circumstances (Huber & McDaniel, 1986).

Decision process management viewed from the two perspectives. The decision-making view of management focuses on those aspects of decision-process management that could be handled with a decision support system; i.e., problem structuring, modeling, and making the decision.[4] Information gathering occurs during the analysis and construction phases of development of decision support systems, and generally means identifying and incorporating the (usually fixed) set of information sources. The other activities would be considered outside the scope of system development.

The information-processing perspective covers basically the same set of decision-process management activities. No one type of system or approach is suggested; instead, there is a range of options. One option might be to use a computer conferencing tool to localize the discussion on a given set of issues, take votes on decisions, etc. Another option might be to improve the manager's information gathering and dissemination abilities through sophisticated filtering, retrieval, and display techniques, and access to a wide variety of communications media.

---

3   This includes communicating the partial results, or the "state of the process," to participants coming in "in the middle."

4   See Meador, Guyote, & Rosenfeld (1986) for a good example of current DSS methodology.

Neither perspective covers the whole process, recognizes it as a process, or suggests how a support system might support this or any other organizational process. Furthermore, neither perspective identifies any common thread behind the activities in decision-process management, taken as a representative process.

Augmenting the decision-making and information-processing perspectives. To augment the decision-making and information-processing perspectives of management work, we restate our earlier contention that the challenge to managers in post-industrial organizations is to sustainably manage the increasing complexity with which they must deal, and we observe that structure is a means by which complexity can be and has been managed (see Chandler, 1962, for historical examples, Galbraith, 1973, and Simon, 1969, for organization design principles). IT can provide help in structuring information (data, knowledge), processes, and organizations (as hierarchies, heterarchies, or markets; Malone, Yates, & Benjamin, 1987). Both the decision-making and information-processing perspectives, and systems stemming from those perspectives, focus on information structuring in one form or another (e.g., models, issue-oriented computer conferences, databases).

We observe that managers engage in a characteristic pattern of action, namely, create, communicate, review, and react (Figure 7.1). Creation is the creation of structure: organization structure, process structure, commitment structure (Winograd & Flores, 1986), problem structure (Humphreys & Berkeley, 1983)— in a phrase, work structure. Communicating means providing information to the appropriate people in order for them to carry out their responsibilities within the structures. Review is the process of monitoring the execution, unfolding, or completion of the structures put in place. Reaction is responding to changes in the environment which threaten the achievement of the goals for which the structures were created.

This activity cycle includes the critical activity of coordination. Coordination means acting to cause different people or groups to achieve some goal to which each one contributes, and encompasses activities such as negotiated goal-setting, direct supervision, and planning.[5] Communicating, reviewing, and reacting, taken together, comprise coordination.

Our IT perspective on management work, then, is that managers create and manipulate work structures in order to deal with complexity. While people are good at purposeful action, they are generally poor at remembering or relating details of many independent or interrelated work structures and multiple instantiations of the "create, coordinate, review, react" action cycle. Computer-based systems, on the other hand, have proven to be exceedingly good at maintaining complex work

5   Note that less obviously directive activities such as education, training, and cultural indoctrination can be considered mechanisms of coordination (Mintzberg, 1979, chapter 6).

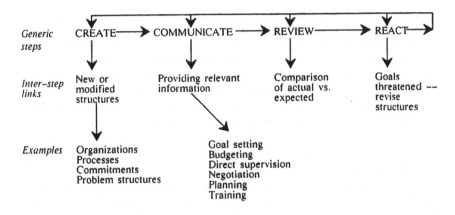

Figure 7.1. The "create, communicate, review, react" action cycle.

structures. Consequently, management support systems must make such structures as explicit as managers wish.

SYSTEM IMPLICATIONS OF THE HOLISTIC PERSPECTIVE

Fundamental Support Requirements for Management Work

The foregoing analysis of activities involved in decision-process management shows the critical importance of four interrelated, fundamental requirements for IT support for managers, all of which are poorly met by contemporary systems.

First, IT must support *group work* and *group processes*. Much of today's IT support is directed toward individuals in the form of tools such as word processors, spreadsheets, and graphics packages, which provide little support for participation in group processes. For example, a word processor allows one person to compose or edit text. Even if it were on a networked computer, it does not provide support for the range of activities involved in two people jointly authoring a document, such as person A commenting on B's work, or allowing B to view A's suggested changes side-by-side with B's own, or allowing A to see the history of changes to a particular section (possibly reverting to an earlier version), or A's influencing which of B's assigned sections B gives priority to doing.

Second, IT support for work (managerial or otherwise) must appear as a *system*, not as separate, unintegrated tools. This means that the user sees a consistent interface to the system's functionality -- identical operations are invoked identically, system behavior is predictable, the behavior of one subsystem does not inter-fere in an obscure way with another, etc.

Third, IT must *flexibly support* patterns of work. To the extent that IT present-ly supports group work, it is usually in two forms: database systems and applica-tions built upon them, and communication systems, usually electronic mail, with computer conferencing a distant second. These systems are inflexible, in that the group processes that they implement (usually implicitly) cannot easily be changed without considerable reprogramming. Database applications such as funds trans-fer, securities trading, ticket reservations and the like implicitly recognize a limited number of different organizational roles, with the number and types of roles being fixed at the time of system specification. Adding a new role is a task for program-mers, not system users. Mail systems, on the other hand, are inflexible in that they provide virtually no process support whatever, nor can they easily be made to (see Holt & Cashman, 1981, for an example).

Fourth, support for *communication, coordination,* and *action* is essential for the process to work efficiently and effectively. The need for communication sup-port is obvious, but present-day IT support for communication falls short in two ways: by generally failing to distinguish the different shades of electronic and hardcopy transmission means by which people can communicate, and by complete-ly separating communicating from working.[6] Coordination is even more problematic. Coordination, in present-day computer-based systems, means coor-dination of readers and writers to avoid various concurrency problems (Bernstein & Goodman, 1981), and even at that level database systems have some technical inadequacies when supporting cooperative work (Greif & Sarin, 1986). Anything above that level is usually left to people to worry about. Computer support for ac-tion (defined in one dictionary as "the transmission of energy, force, or influence") is widely provided, but within the limitations of the inflexibility of present tools.

Our focusing on IT to support communication, coordination, and action may seem somewhat at odds with the prevailing emphasis on decision-support systems to aid managerial decision making. We believe that current individual and group decision-support systems have targeted the information processing aspects of decision making while largely ignoring the matrix of other activities in which decision making is embedded (Kraemer & King, 1986). Our analysis of decision-process management emphasizes the range of activities involved in "decision making," and hence the need for IT support for the full range of the activities. Fur-thermore, as complexity and turbulence increase, the need for effective support for communication, coordination, and action will increase.

6   That is, in order to communicate, one leaves the current task (e.g., an invocation of some tool) and starts up another task (a communication tool); the communication does not occur as a natural consequence of the task performance (i.e., the tool use).

System Design Implications

Our structural perspective implies that the structure of information, processes, and organizations must be made explicit and kept separate from content. To understand this, we must examine the classes of computer-based systems that people have constructed to explicitly manage these kinds of structures, and for this purpose we recapitulate the framework developed by Cashman and Stroll (1987) for analyzing how IT can support cooperative work. This framework is relevant because many cooperative work support systems depend on explicit representations of information, process, and organization structures.

Classes of computer systems for supporting cooperative work. Malone's (1985) concept of an organizational interface is the organizing concept for our understanding of how IT can support cooperation. He observed that computer systems were being built in which relations between people were being mediated by computer connections, and he called these computer-mediated human relations the organizational interface of the computer system. He singled out the organizational interface as a subject for study and design in its own right, bringing in ideas from organization theory and economics to apply to the design task.

To analyze the area of IT support for cooperative work, we apply the idea of the organizational interface in two ways. First, consider the formality of the organizational interface. A *high formality* interface is highly governed by rules, procedures, forms, and restrictions on who can talk to whom at what time, about what subject, and in what format. A *low formality* interface is marked by ad hoc interactions free from any restrictions. The property of formality is really a property of an organization or any part of it, and has nothing to do with IT support per se. It will be seen that not only does formality differ from one organization to another, but that different parts of the same organization differ widely in formality.

The second property relates directly to IT support, namely, the flexibility with which the formality of the organizational interface can be adjusted. An IT support system with *high flexibility* can easily accommodate organizational interfaces with a wide range of formality within the same system; moreover, a particular interface may start out low in formality, but be increased in formality as the participants identify aspects of their interactions that can be rationalized. (The reverse situation would also be accommodated in a high flexibility system.) In contrast, an IT system of *low flexibility* can accommodate high formality interfaces or low formality interfaces, but it is difficult for people other than the system's programmers to change the interface to modify its formality.

The two properties, each with its two poles, lead to four quadrants as shown in Figure 7.2. We call systems inhabiting the high formality/low flexibility quadrant *office procedure systems*, since historically the systems in this quadrant were oriented to fairly routine office work. Examples of such systems are MONSTR

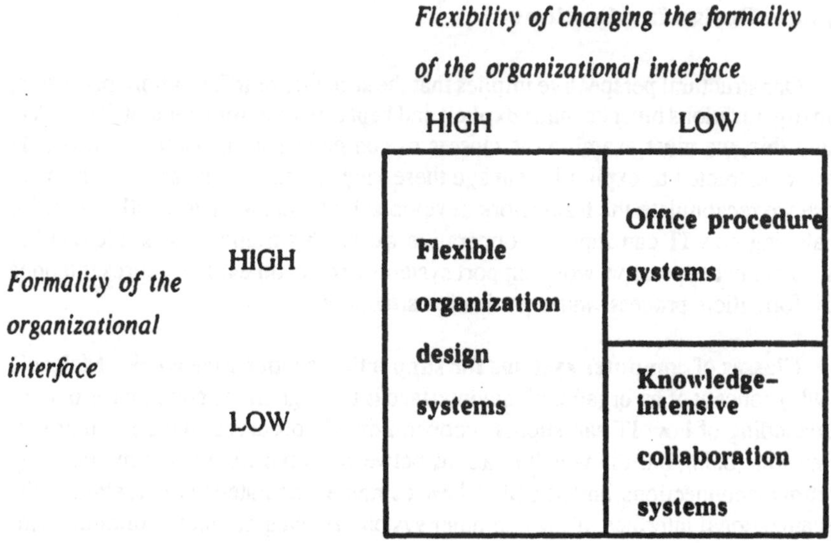

Flexibility of changing the formailty

of the organizational interface

HIGH   LOW

Flexible organization design systems	**Office procedure systems**
	Knowledge–intensive collaboration systems

Formality of the   HIGH

organizational

interface

LOW

Figure 7.2 A framework for classifying computer systems to support cooperative work.

(Holt & Cashman, 1981), XCP (Sluizer & Cashman, 1985), SCOOP (Zisman, 1977), CMS (Kedzierski, 1983), and POISE (Croft & Lefkowitz, 1984).

   Knowledge-intensive collaboration systems inhabit the low formality/low flexibility quadrant, and are so called because they generally support interactions in which exchange of knowledge or information is done in ways not governed by strict organizational procedures. Systems in this quadrant include electronic mail systems,[7] computer and video conferencing systems, real-time shared-screen conferencing systems (Sarin & Greif, 1985), meeting augmentation systems such as Colab (Stefik et al., 1987) and Project NICK (Begeman et al., 1986), and hypertext/cooperative authoring systems[8] such as Augment (Engelbart, 1984a, 1984b; Engelbart, Watson, & Norton, 1973), Intermedia (Garrett, Smith, & Meyrowitz, 1986), and NoteCards (Trigg, Suchman, & Halasz, 1986).

   The essence of a high flexibility system is that organizational interfaces of a wide range of formality can be supported within a single system. The distinction

7   It is interesting to note that the MONSTR system grew out of the inability of a mail system to support an application that required a high formality interface.

8   Hypertext is a form of computer-stored text in which pointers from one segment of text to another are embedded in the text. A hypertext system (which can include graphics and other media besides text) provides facilities for "jumping" from one text segment to another via these pointers (or "links"), creating and deleting pointers, etc. See Conklin (1987) for a survey of hypertext systems.

between the low and high formality quadrants in this column effectively disappears. Therefore, we refer to systems in this high flexibility column as *flexible organization design systems*, because their purpose is to support the design and evolution of electronic organizations. Soma (Holt & Ramsey, 1985), CHAOS-1 (De Cindio, De Michelis, Simone, Vassallo, & Zanaboni, 1986), and the Information Lens (Malone, Grant. Lai, Rao, & Rosenblitt, 1986) are research prototypes of this class of system.

   Representing information, process, and organization structure. To survey the means by which designers have represented information structure generally is beyond the scope of this paper. For the purposes of supporting management work, we note that explicit representation of document structure is a reasonable place to begin. This would include document organization into chapters and smaller units, inter-document linkages such as citations or the representation of debates based on the information contained therein (Lowe, 1986), document change history (represented so as to be able to reconstruct earlier versions), alternative presentation formats (sometimes referred to as a compound document architecture), etc.

   Work on representing organizational processes began with the early work on office procedure systems (Ellis, 1979; Marca & Cashman, 1985; Marca, Schwartz, & Casaday, 1987; Zisman, 1977). The representational scheme of the Soma system (Holt, 1986; Holt & Ramsey, 1985) is a good example of what must be explicitly represented. In this formulation, a person's electronic work environment is composed of a set of *centers*, where each center represents a function for which the person is responsible. Within a center are the *objects* that form the work materials for carrying out the function of that center. These objects may be permanently located at the center or may be transient. Also contained within the center are the *roles* that the person plays when carrying out the center's function. Each role has a *role script* that describes the behavior of the role, such as what a person playing the role can or cannot do; what other roles the role player can interact with, and in what manner; and what objects the role can manipulate. Centers and roles have connections to other centers and roles. The behavior of the connections is governed by the role scripts of the roles that are connected.

WORK IN PROGRESS TOWARD A MANAGEMENT SYSTEM

Background: The NB Business

   We are in the early stages of developing a prototype management system to support the work of an organization, most definitely including the managers, engaged in developing a business within Digital Equipment Corporation. (We refer to this business simply as "NB.") This effort has three goals: to support the work of the organization as well as possible, to help clarify the requirements for and attributes of a Management System, and to gauge the impact of the introduction of a system of this kind on the members of the organization.

For all intents and purposes, the environment in which the NB business exists is a post-industrial environment with the characteristics described by Huber. Three aspects of the NB business make it particularly complex. First, there is *product complexity*. The NB business is to produce entire application systems, including hardware, system software, and application software for a diverse, highly competitive marketplace. System performance is critical, and there is no single measure of that performance. Although the company has considerable experience with complex components and products, it has less experience developing the requirements for, specifying, and building application systems as complex as this.

Second, there is *process complexity*. The product and market complexity made it necessary to develop a new approach to systematically relating customer requirements to system metrics to which engineers can design, to the training salespeople must receive, and to the capabilities that other Digital organizations must put in place. This is developing into an entire *business process architecture*, with a number of interlinked, often novel processes for formulating, refining, communicating, and validating requirements for systems, component products, marketing and training programs, etc.

Third, *cultural change* will be required to deliver complex application systems developed via a novel, information-intensive process. Specifically, the NB Marketing/Engineering group is taking a very active role in developing customer requirements and understanding what comprises value to the prospective customers. Rather than being technology-driven, the group is striving for a balance from the earliest stages of product definition. Another aspect of cultural change is that many of the new hires in the NB marketing group are new to Digital, and must acclimate themselves to its culture.

Flexible Organization Design System

We plan to implement a flexible organization design system as described previously. We will test its ability to flexibly support evolving processes by prototyping a program management application for the NB business. This application will assist the NB program manager and the "line of business" managers (each targeted at a significant market segment) in tracking the goals and deliverables of the latter. Program management, in this context, includes, but is not limited to, activities such as developing the format for program reviews; holding program reviews; tracking action items coming out of reviews; negotiating commitments between program managers and team members; tracking and renegotiating commitments; scanning the program activities and interdependencies for actual or potential "holes" that must be attended to; evaluating program activities and interdependencies against changes in the environment; planning, executing, and monitoring actions to correct a problem or seize an opportunity; and so on.

Structuring and Retrieval Mechanisms

We are just at the requirements definition stage for these capabilities. In the interests of providing an information retrieval capability for the NB management team now, we are installing a document database system that offers keyword and full-text retrieval of documents stored electronically, and keyword retrieval for paper documents that have been catalogued. The capabilities of this tool are program-callable so that it can be invoked by the role scripts of the flexible organization design system.

As part of our prototyping effort, we are exploiting this retrieval facility to be aware of different levels of abstraction of the source material being retrieved, as well as its structure. For example, a presentation may consist of a set of slides, a report, or both. A slide set consists of a number of individual slides, whereas a report has sections, subsections, subsubsections, and paragraphs. Furthermore, any item may refer to one or more other items, where the reference implies or explicitly states some relationship (e.g., whole/part relationships, version relationships, bibliographic references). Users will be able to exploit these structural relationships by including or excluding certain relationships from participating in a retrieval.

Modeless Operation

In current computer-based systems, users see individual tools—the text editor, the mail system, the spreadsheet package. Each has its own command language and conventions. A user task, such as developing a proposal, may encompass the use of many tools, each invoked a number of times, with files being moved back and forth from one tool to the next. This will be replaced by *modeless operation*, in which the user activates tasks which are meaningful to him or her—develop a proposal, review the progress of the X business, schedule Y's performance review. Each user task may cause the invocation of tools, but these invocations will be invisible to the user. Modeless operation will be implemented through extensive use of the role-script facility of the underlying flexible organization design system.

MANAGEMENT SYSTEMS IN COMPLEX ENVIRONMENTS

To gauge the likely effects of a management system on managers in post-industrial organizations, consider the decision process management example described earlier. The management system will enable many constituent activities to be performed in ways that IT at present barely supports.

First, it will be possible to create processes that describe the roles involved in the process, the actions that can be performed by an agent (person or machine) in a role at any stage of the process, the objects that are inputs and outputs to the actions, and the conditions that the objects' states must satisfy for the actions to take place. This capability will allow people to describe a decision-management process as a process that the computer can now interpret and support. Subsidiary processes can be designed to fit into higher-level mandated processes. Also, generic processes such as decision-process management can be defined for a company, division, or smaller group, and can be tailored to fit particular circumstances. The other side of this structuring is that it requires a much higher level of discipline in the organization, and the people must perceive that the benefits of the discipline exceed the extra effort to create and maintain the structure.

Second, management systems will enable people to modify processes quickly and cheaply, preserving organizational flexibility in the face of turbulence.

Third, management systems will enable people to execute and monitor these processes in a distributed fashion. This capability insures that decentralized organizations can continue to operate in that manner. Also, by dividing processes by the roles involved, a person's electronic world can be monitored very quickly for roles where the person's attention and intervention are needed.

Fourth, as decision support systems are built on the foundation of the management system, they will better integrate decision making with communication, coordination, and action.

The net effects for managers are:

- Coordination can be done electronically in a way which is not now possible
- Larger teams of people can be coordinated for longer periods of time, and across unlimited distances.
- More complex processes can be coordinated, because:
-
    - They can be specified and agreed upon, or alternatively, they can evolve organically.
    - They can be examined, since they are in an explicit form.
    - They can be executed in a distributed fashion and can be monitored "in real time," if so desired.

- The manager's effectiveness and efficiency are increased, because:
  - The management system can bring to his or her attention those roles or processes requiring intervention or attention.
  - The manager can decide which items requiring attention are critical, and can go right to those, bypassing non-critical items.
  - More work can be done (due to increased complexity which is now manageable).
  - Higher quality work can be done (more specialists can be brought into the loop, if necessary; less missed chances due to improved coordination; faster handling of tasks due to modeless operation).

SUMMARY

Today's business environment is complex and becoming more so, and managers must try to sustainably manage this increasing complexity. Information technology (IT) can help managers achieve this end by providing powerful means to structure information, processes, and organizations. An examination of the numerous small constituent activities involved in organizational decision making reveals that IT must flexibly support group work and group processes, which in turn means that facilities for communication, coordination, and action must be provided. This represents a key shift from the present thrust of IT support, which is oriented toward individual information processing.

At Digital Equipment Corporation, we are in the early stages of developing a prototype management system. Our goals are to support the work of a new business group, to better understand the requirements for and attributes of a management system, and to gauge the impact of such a system on its users. The foundation of the prototype will be a flexible organization design system now under development, and at present the system provides a prototype tool for structuring and retrieving information stored in a variety of types of repositories located on a local or wide area network.

Management systems of the type we are developing will allow managers to effect coordination in ways not presently possible. Larger teams of people will be able to engage in more complex processes for longer periods of time, and across unlimited distances. A manager's effectiveness will increase, because the organization of the system will enable items requiring critical attention to be brought more easily to the manager's attention. The manager's efficiency will be enhanced, because more complexity can be harnessed, and higher quality work can be done.

ACKNOWLEDGMENT

The authors wish to thank Dick Dobbie for his helpful comments on this chapter.

8

TOWARD PORTABLE IDEAS

Mark Stefik
John Seely Brown
Xerox Palo Alto Research Center

INTRODUCTION

We honor creativity in our culture, especially that of the individual genius, but creativity is as much a social as an individual affair. When people of different backgrounds come together, new ideas can arise from their conversations. Sometimes new ideas are built up incrementally from the fragments of different viewpoints. Ideas can be made more robust when they have been bounced around, critiqued, polished, and repackaged by a group.

We have all been in situations where our part in developing ideas has been but one of many contributions. We have all benefited from the wisdom of a second opinion, or been surprised on occasions when good ideas came from unexpected sources. Nonetheless, idea creation, like other aspects of intellectual work and even routine office work, is usually conceived in terms of the contributions of isolated individuals. Conventional wisdom has been slow to recognize the importance of collaboration and teamwork. The wisdom about teams that "many hands make light work" refers to hands not minds, and certainly not to committees.

Most approaches to office automation and computer-mediated work have focused on individuals rather than groups. The landscape is littered with failed computer systems that were supposed to make light work of various office tasks (Bikson, 1987). Studies of the acceptance and use of computer systems in offices have shown consistently that a major factor determining the success or failure of such systems is whether the designers took into account the habits, needs and activities of work groups.

During the past two or three years, however, there has been a burst of new thinking about computer-mediated work. In contrast with the terminology of personal computers and computers to empower *individuals* to do their best, we are starting to hear much more about *interpersonal computing*. This means different things to different people, and new jargon has started to appear including "coopera-tive computing," "collaboration technology," and "work group computing." The somewhat awkward term "groupware" has been proposed for computer software that is specifically intended to aid in the work and coordination of activity of a work group.

This chapter presents a particular vision of possibilities that we have found intriguing and that we believe could have profound effects on the functioning of organizations. Crafting tools that actually help collaboration is a very subtle enterprise. There are two parts to our thesis. The first is that creative genius lies in the social substrate itself. Secondly, the interaction of ideas properly externalized and appreciated leads to wonderful combinations and results. Some of this syner-gy and exposure can be enhanced by tools in the social infrastructure. There is a streak of genius and creativity in each of us. That streak can be tapped by creat-ing a medium in which ideas can rub productively against each other. We propose a medium of active and sharable workspaces for developing and explaining infor-mation.

In the next section we focus on the Colab project, one of several projects study-ing collaboration and supporting technology at Xerox PARC. We sketch out the basic premises of the project, describe one of the experimental systems that has been developed, and present some of the questions and issues that we have en-countered in the work so far. In the following sections we reexamine and critique some of the basic assumptions of this work. This prepares us to ask more basic questions, and also to propose ways that the ideas and technology for group work could prove more significant and valuable than in our original vision, especially for research and engineering organizations.

The study of collaborative tools requires transcending what we call the "tech-no macho" syndrome, the fascination with technology or methodology for its own sake. We should not get carried away with the belief that technological artifacts or decision methods in themselves will help that much. Many obvious attempts to apply technology to the work setting have hindered more than they have helped. However, for the authors, there is no denying or escaping our role as technologists. We are not disinterested observers of technology and the social scene. We are, in fact, concerned with the limitations of the status quo and are actively trying to in-vent new and more productive ways of working. This chapter is intended to stimu-late those who want to think beyond current technologies and work practices.

COLAB MEETINGS AND CONVERSATIONS

The Colab project (Stefik et al., 1987) was conceived as an experiment in com-puter support for meetings. We imagined that professional people in meetings should have the same kind of access to computers that they have in their offices for private or isolated work. In support of this we created the Colab meeting room as shown in Figure 8.1. The meeting room provides a computer workstation for each participant in a face-to-face meeting. At the front of the room is a blackboard-sized touch sensitive screen (which we call a "liveboard") capable of displaying an image of approximately a million pixels.

Most of the meetings that we had in mind when we started the Colab project take place by a small group in front of a whiteboard or some other vertical and erasable writing surface. In these meetings, a creative group is engaged in discussion and work activities using notations on the whiteboard to formulate and explain their thoughts and to keep notes during the meetings. For us, the whiteboard is the dominant medium used in meetings; it is a medium that we all use constantly in our daily work. We could see many shortcomings of the whiteboard, several of which are discussed in the following. The whiteboard became the technology to beat in inventing a more powerful medium for meetings, and we decided to beat it by creating a computational medium that kept the best properties and brought in new capabilities as well.

Our new medium distributed a computational whiteboard to every participant in a meeting. To promote shared viewing and shared access to what is written during the meeting, the Colab software is oriented around a concept for multi-user interfaces that we call WYSIWIS (What You See Is What I See—pronounced "whizzy whiz"). In a WYSIWIS interface, all the meeting participants can see exactly the same information on their displays. Colab meeting tools support this illusion by maintaining synchronized views across workstations. In addition, each person can point to things on the display with a personalized "telepointer" that is made visible in real time to the other participants. Colab software also supports private windows. Private windows correspond to notepads; public WYSIWIS windows correspond to whiteboards.

But how could computers possibly help in meetings? One could approach this question by enumerating the aspects of meetings that are annoying and then investigating which of them might be ameliorated by computer technology. This would be an awesome task, requiring at the onset some substantial focus to limit and to identify the kinds of work to be investigated. However, we approached the issue from another direction, trying to understand the properties of a computer medium and then imagining the kinds of meeting situations in which computers could make a positive difference. Most of our intuitions are based on information processing concepts both for computers and the nature of work in meetings. These concepts also provide some insights about the kinds of work activities for which computers might make a difference.

Computers provide more space for writing than whiteboards. The storage capacity of a whiteboard is quite limited and after a period of time, a group writing on a whiteboard must erase things in order to keep going. In a computer medium, the display space can be reused without discarding information because

Figure 8.1. View of the Colab. The Colab is an experimental meeting room that provides computational support for collaboration in face-to-face meetings. It is designed for typical use by two to six persons. Each person has a workstation connected to a personal computer. The computers are linked together over a local area network (Ethernet) that supports a distributed database. In addition to the workstations, the room is equipped with a large touch-sensitive screen and a stand-up keyboard. (Photograph by Brian Tramontana)

symbols can be moved to and from file space. Furthermore, the file storage capacity of a computer is quite large and there are many techniques for organizing the display of information on a computer screen using windows, icons, and scrolling techniques.

Even with whiteboards, participants tend to build up large collections of written symbol structures that provide the common ground for reference. With more space for writing, participants can build up potentially larger sets of shared writings. This can be important for meetings that last for several sessions. Backup on a file system makes it possible to recall and display things even if they were developed in previous meetings.

The abundance of space in a computer is no excuse for neglecting to manage space as a resource, but here again the computer medium offers some advantages.

Using techniques from bitmapped user interfaces, items can be quickly and easily rearranged on a computer screen. In contrast, on a whiteboard one must manually copy and erase symbols in order to rearrange them. This flexibility makes it possible to organize a screen, reducing clutter. This enables a group to organize space more easily for the purposes of visualization, and accommodation of shifts in focus.

A computer medium provides computational leverage that can be used in many ways. For example, in a resource allocation meeting, the computer could provide visible spreadsheet capabilities. It could display information in alternative formats for easier manipulation or better understanding. It can also provide search services for finding information in large sets.

WYSIWIS interfaces can relax some constraints on communication and cognitive processing. By enabling participants to use a shared written medium, the bandwidth of communication is potentially increased since more than one person can add information at the same time. However, even if the apparent increase in bandwidth of communication in a Colab setting is not significant, freeing the constraints on parallelism and serial communication may improve the quality of deliberations by enabling meeting participants more freedom in scheduling their attention and cognitive activities.

These general capabilities of a computer medium suggest that the Colab would have the most advantages in meetings that include manipulation of substantial amounts of information, such as meetings by engineers in which complex designs are discussed and compared.

Meetings Tools: An Example

We use the term meeting tool to refer to computer software in support of groups in meetings. Just as users of personal computer software need different tools for different purposes (e.g., text editors, spreadsheets, mail systems), so too do meeting participants need different tools for different purposes (e.g., tools for agenda control, brainstorming, negotiation, and argumentation).

Cognoter is a Colab meeting tool used in our lab about once a week. It is used to organize ideas for presentations, reports, talks, and papers. Cognoter supports a meeting process in which participants come together, usually without having prepared any materials ahead of time. Meeting participants determine the audience and goals for their presentation, the topics to be included, and the overall organization. The output of Cognoter is an annotated outline. Figure 8.2 shows an example of the visual display created by Cognoter when it is being used.

The organization of Cognoter is described in the following. To convey a clearer sense not only of what it is but also how it is used, we have included some informal observations of use and meeting phenomena. However, at the time of this writing, adequate recording capabilities for observing meetings were just becoming operational. Consequently, formal and quantitative studies of group behavior in the Colab are still ahead of us.

Several things can reduce the quality of a presentation. It could fail to include some important topics; it could dwell on irrelevant or unimportant topics; or it could address the topics in an incoherent order. To avoid these pitfalls Cognoter organizes the process into specific stages. Each stage incrementally increases the set of actions available to the user.

The stages in Cognoter are brainstorming, ordering and grouping, evaluating, and generating an outline. We originally adopted this structure from a similar one that proved useful for us in non-computational settings; however, there are significant differences in the uses and effects of the stages when computers are introduced.

Brainstorming Ideas. The brainstorming stage is intended to foster the free-flowing contribution of ideas. There is one basic operation: A participant selects an empty space in a public window and types in a word or phrase characterizing an idea.

Unlike usual brainstorming meetings, there is no waiting for turns in Cognoter; any participant can enter an item at any time. Often the inspiration for an item is triggered by another participant saying something or entering an item in a public window. Thus, communication (or loosely "the conversation") in Cognoter takes place both by voice and over the computer medium. All items appear on everyone's displays. Participants can annotate items with longer descriptions to clarify their meanings.

Organizing Ideas. The order of items for the presentation is established in Cognoter by incremental and local steps. There are two operations: linking items into presentation order, and putting items into delineated subgroups. If item A is linked to B (meaning A comes before B), and B is linked to C, then A comes before C. If item A is linked to a subgroup, then it comes before every item in the group. By these transitive and distributive operations, a small number of explicit links can tightly constrain the order of items in the outline.

The linking operation often takes place in conjunction with an oral justification. For example, if "expenses" and "bottom line results" were items, a participant might argue out loud "We have to talk about expenses before bottom line results because otherwise management won't understand the results." This relation is represented visually in Cognoter as an arrow linking the item labeled "expenses" to the one labeled "bottom line results." It is also possible to move related ideas to an idea subgroup in a separate window. Before moving items, it is common practice to put them in a spatially compact cluster. This allows comment on the coherence of the proposed grouping.

In Cognoter, the overall task is richer than in traditional brainstorming sessions. For one thing, the task is not finished when some ideas have been generated. Preparing a presentation requires organizing and evaluating ideas as well. Further-

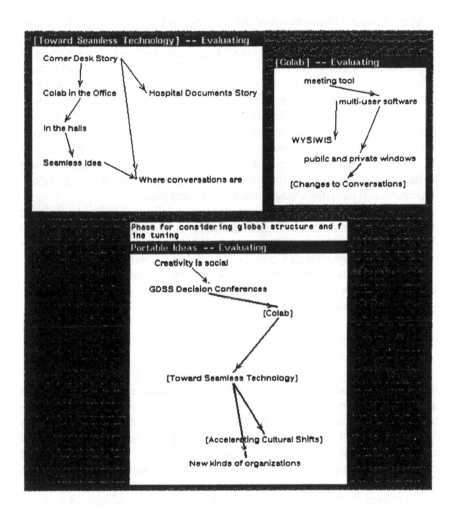

Figure 8.2. Interacting with Cognoter. Cognoter is a meeting tool for organizing a presentation. Cognoter guides this process through several stages: brainstorming, organization, evaluation, and outline generation. The items on the display are short expressions that refer to the ideas for the presentation. In this figure the items have been organized into three major groups. The arrows between the items indicate an ordering relationship; they imply constraints about items that must be presented before other items. Cognoter automatically generates presentation outlines that take into account these constraints.

more, our informal observations of the meetings indicate that people form subgroups that focus on the development of particular aspects of the subject matter. Since subgroups of ideas are usually put into separate windows, each subgroup of people can focus around one window or another. The frequency and significance of this behavior and the importance of supporting it with meeting tools will be a subject for systematic study when our observational facilities become operational.

Although subgroups of collaborators can work mostly independently, they can also communicate. For example, one group may decide that some of the items in its windows don't fit with the others, and may put them back into the general pool or offer them to another subgroup. Further communication is then required when the items are reconsidered, perhaps by the whole group. When subgroups rejoin, participants can recap the changes made in the subgroups.

Evaluating Ideas. During this stage the subgroup boundaries tend to dissolve and the meeting participants function again as a single group. Participants try to understand the organization of the presentation as a whole. Items that seem irrelevant or less important than others can be deleted. Outlines are generated by Cognoter upon request, and ambiguities in the ordering can be highlighted. Participants can argue whether particular items are irrelevant or unimportant when compared with others.

Expanded Dimensions for Conversations

Tools like Cognoter embody more than a shift from a whiteboard to a computer. Effective collaboration has some tacitly held rules, such as taking turns in conversation. There are multiple roles such as inventing, critiquing, reformulating, scribing, and summarizing. People switch roles during a conversation, and the switching itself follows certain rules bearing on rhythm, momentum, and topical focus. For example, in an effective collaboration one party will hold back and not interrupt a second party who is obviously "on a roll" generating a stream of related ideas.

Computer media change some of the basic parameters of conversation and enable profound changes in the shape of the conversations. When we move from personal to interpersonal, the requirement for personal intelligibility of the subject matter shifts to a requirement for *mutual intelligibility;* the meaning of conversational terms shifts from being internalized and fixed to being externalized and negotiated. Expressed in a computer medium, communications persist in a form that is tangible, external and manipulable. We conjecture that this substantially increases the amount of information for which there is a lasting awareness and shared understanding by participants in a meeting.

The coordination of intellectual work around manipulable icons draws on familiar skills for the coordination of physical work. Sorting can be done with icon manipulation. One moves items between buckets until they are in the right places.

In both cases the multi-user interfaces indicate when objects are being manipulated (e.g., edited) by someone, providing visual clues for team coordination.

The same kind of manipulative action can be used for indicating when one participant wants another to work on an item. Thus, it is possible to pick up an item and drop it into the workspace of a second participant. This is very much like the physical act of picking up a physical object (e.g., a piece of paper) and handing it off to someone else for attention. A single action removes the object from one's own inventory and adds it to someone else's inventory for attention.

The possibility of simultaneous communications relaxes many constraints. It increases the possible bandwidth of communication in meetings. Changes like this raise many questions for which we do not yet have answers. In ordinary conversation, the meaning of what is said often crucially depends on the context of what was said just before. In Cognoter, multiple things can be "said" at once in the computational workspace. It can be argued that such capabilities introduce confusion into the meeting process. On the other hand, this capability can be enormously freeing in the context of a fast-moving brainstorming session. Reading is faster than listening so it is possible to scan the items being created by several others and occasionally to respond to them. When something puzzling comes along on the screen, however, it is not necessary to tend to it at once. Unlike oral communication, there is no need to remember a confusing item because it remains in the workspace inventory for later processing. Any systematic process for going through the items will encounter the item again for later consideration. A written workspace is amenable to scheduled and systematic processing of comunications. Thus, even if the potential increase in bandwidth does not result in an overall increase in communication, it may be important for other reasons.

During the brainstorming phase of Cognoter, the parallel action in proposing ideas reduces the usual verbal communication to coordinate turn-taking and synchronization. A participant can enter an idea whenever it comes to mind. Oral conversation tends to drop off radically, since so much of the communication load shifts from ears to eyes. The speech resource becomes more available for questions that clarify points.

It is important to note that parallel action is not altogether absent in non-computational meetings. Videotapes of design meetings show that groups of designers working on large sheets of paper engage in much parallel sketching activity. Furthermore, shared workspaces can be created in other media such as video. Architects working together through such a medium have reported an intensity of engagement and productivity (System Concepts Laboratory, 1987) similar to the informal reports by users of shared workspaces in the Colab.

Another profound change to the dimensions of conversation is the possibility of equal access to public data. In Colab, the conversational acts that enable a participant to modify the public display or to assume the role of chairman can take place in a fraction of a second. In contrast, with an ordinary meeting room with table and blackboard one must negotiate the transition, rise from a chair, walk to the board and so on. By lowering the hurdles of transition the technology creates a potential for broader participation and for more flexibility in roles.

Whether and when these changes in conversations are beneficial is still to be determined. For example, accelerating the pace of a meeting as in the brainstorming phase may give participants less time to think. On the other hand, freedom from serial turn-taking may alleviate some of the problems of production blocking (due to the limitation that only one group member can talk at a time) reported by other studies of brainstorming in more conventional media (Diehl & Stroebe, 1987). In some cases, the context surrounding the generation of an item may be lost. Important effects also happen at the transitions between meeting processes. Among the transitions are the formation and dissolution of subgroups, the shifting of participants from one topical focus to another, and the transitions of the conversational patterns of the whole group as it shifts from having a single focus of activity to having multiple conversations and subgroups and then back again. However, our purpose now is to begin to understand some of the differences, not to evaluate them.

Seamlessness at Work

As we move on to the next phases of our research, we are also ready to challenge the assumptions of the Colab design with an issue that seems to dwarf all the others:

Meetings do not take place (exclusively) in conference rooms.

Meetings take place wherever people get together and have conversations. Recently, one of us spent time observing the use of documents by nurses at the Pacific Medical Center in San Francisco. What was impressive and interesting was how much clarification, co-ordination, and negotiation among nurses took place over their clipboards. If a narrow-minded computersmith wanted to bring information processing to the hospital situation, the first bad idea might be to make the documents available on a workstation located somewhere down the hospital corridor. That would completely ignore the conversations and interactions of the nurses where they meet. Like other intelligent human beings, nurses could probably cope with a poorly designed computer system. However, to be most helpful, we suspect that the technology of record keeping and conversation should be as familiar and easy to use as a blackboard and as readily available and portable as the clipboard, paper, and pen.

Figure 8.3. Team programming at a corner desk. Wedge-shaped desks were developed for use by our group in team programming. A critical factor bearing on the successful use of these desks was locating them in offices. Programming activities start in offices, and teamwork begins there. When we first placed the desks in conference rooms, we observed that programming that started in an office never moved out to the desks. Instead, collaborating programmers would just squeeze into the office situation. The problem was presumably that the overhead of moving the computing environment and debugging context to a second location was greater than the benefit of increased elbow room. (Photograph by Brian Tramontana.)

To illustrate this issue of the use of technology and the location of meetings, we present a story about some very low technology that we misapplied at PARC. The technology was a corner desk, a wedge-shaped desk for holding a computer workstation that could be located in the corner of a room as shown in Figure 8.3. We wanted to promote team programming on our research projects and these new desks offered ample elbow room for two people sitting together. We placed the desks in our regular conference rooms, equipped them with computer workstations, and anticipated that they would also be useful for demonstrations and for visitors when office space was tight.

After the corner desks were installed, we noticed that they were never used. Team programming was happening occasionally, but it always started in somebody's office. Someone might start with a system debgging question, or have a programming puzzle or idea. In every case people jammed their chairs together in the office and squeezed around the workstation. At no point did they move to the corner desks in the conference room.

Concluding that the desks were a failure, we decided to have them stored in a warehouse. To salvage something from them, one of the authors who liked the aesthetics of the corner desks and had an office shape that could use one easily, decided to discard a table and regular desk from his office and to use a corner desk instead. Shortly thereafter, team programming was observed to occur in his office, and furthermore, the idea of putting the desks in offices was legitimized. The corner desks are now mostly located in our offices, where programming occurs. Now they are serving their intended purpose, team programming occurs on a regular basis, and several more corner desks have been built.

Returning to the assumption behind the Colab meeting room, we note again that meetings take place regularly in offices, not just meeting rooms. This raises the question of how offices should be equipped. Here we believe that the Colab experience is relevant. In taking a prescriptive stance, we can predict what could be possible.

We believe that one of the most useful additions to the infrastructure of an office would be a large touch-sensitive display: an office liveboard as sketched in Figure 8.4. A large display creates a focus of attention for a team working together. Furthermore, with other workstations based on CRT displays or flat panels around the room, we could presumably get some of the WYSIWIS meeting phenomena that we have observed in the Colab.

However, there are two main features that intrigue us. One is the power of the computational medium for flexibly organizing space. Whether whiteboards are in conference rooms or offices, they never have enough space. All of the arguments that we advanced for Colab workstations for managing display and file space apply equally well to meetings in an office. The second feature that intrigues us is the possibility of an accurate, large-scale pointing device adequate for quickly sketching diagrams on a high resolution liveboard. The whiteboards of PARC are always filled with informal diagrams and symbols of great variety. The liveboard in the Colab, however, has a resolution of somewhat greater than one pixel per tenth of an inch. Although this is good enough to present a large image of one of our computer screens, it is much too coarse for smooth sketching at the liveboard.

This brings us back to our suggestion that the technology of conversation should be as familiar and easy to use as a whiteboard and as readily available as

Figure 8.4. Redesigning the office. Colab-like devices could lead to a radical redesign of the office environment, in recognition of its use to support small meetings. For example, the environment could integrate personal electronic notepads with public liveboards on the walls. (Line drawing by Steve Osburn.)

paper and pen. Paper, pen, and whiteboards seem natural and easy to use, in part, because we are exposed to them and trained in their use at a very young age. They are also supremely flexible. You can write on them in almost any way that you desire, making text and figures slant up, slant down, surrounded by wiggly globs, and so on.

Many other meeting devices would be possible in an office and would work well with a liveboard, such as multiple keyboards, remote pointing devices, and small flat panel devices used as networked sketchpads. As in the Colab, pointing devices (e.g., mice or styluses) could enable participants to point to something on the public display without leaving their seats. All these devices should be small so that they could be stored out of sight, and they should be cordless. Another useful device would be a digital tape loop and audio gear so that one could recover a particularly apt turn of expression by playing back the last part of a conversation.

The well-equipped office should provide an information environment that con-nects seamlessly larger and more formal meeting rooms like the Colab. The same software and hardware should support conversations in both settings. In an office one should be able to prepare materials for larger meetings, and to continue small-scale follow-up meetings afterward.

Explaining an idea to a colleague is simplified when the context-setting sketches on a whiteboard are available. With active liveboards in offices and user interfaces based on new remote window systems, moving the contents of a liveboard from one office to another will be possible. To this end, means for moving the contents of a liveboard should be direct and simple. One could for-ward the contents to an office liveboard ("Send this to my office") or file it in a database. Similarly, one could retrieve something to show to a colleague in the coffee room ("Get the big-idea window from yesterday's conversation.")

This brings us to a most important concept about new capabilities: the port-able meeting. A meeting that starts at one office on one day could be resumed by any of the participants in their own office or introduced to another colleague at yet another time or location. In conventional meeting situations, different kinds of records are kept for different purposes. Some things are written during the meet-ing for the purposes of explaining or developing a point; other notes are written by the participants for later use by themselves or others, and yet other things may be written such as minutes for explaining to parties that were not present at the meet-ings. With portable meetings the explanatory scribbles created during the course of a meeting become available for reuse at a second location without the need for manual copying. Furthermore, if one person explains a set of ideas using figures and symbols from a liveboard, the second person could gain access to this infor-mation and extend the script for explaining it to a third person.

Many visitors to Xerox's System Sciences Laboratory are surprised by the number of floor-to-ceiling whiteboards, each in its own corner with comfortable seats or couches around them. This is not an accident or just a sign of opulence. These areas were explicitly designed to foster small collaborative teams working in semi-private areas. Whiteboards enable people to create a large sharable con-text. By having so much of the discussion visibly displayed, it is very easy for someone walking by to gain a sense of what is going on and to decide whether to contribute. Further, one can come up to speed more quickly.

The phenomenon of portable meetings may enable additional possibilities if liveboards were introduced into public areas. One of the most heavily and produc-tively used whiteboards in our laboratory is the one near the coffee service in the lounge area. Perhaps coffee centers foster creativity both because people encounter each other here, and because of the informality and relaxation that the centers sug-gest. Could the creative power of a center be tapped better if the whiteboard were replaced by a liveboard? After productive conversation one would not need to remember the ideas or to copy the contents to paper (see Figure 8.5).

Figure 8.5. A well-equipped work group. Colab-like devices could lead to a radical redesign of all the areas where informal meetings take place. In a setting around the coffee pot, one could go up to the liveboard and retrieve a board from a previous conversation, or forward the board contents so that a conversation could be continued in an office. (Line drawing by Steve Osburn.)

This, finally, is what we have in mind for infrastructure for group work: a seamless environment of tools for conversation that extends from offices to the coffee room to the formal meeting room. One might even consider surfacing the tables in the cafeteria with interactive flat displays, providing an electronic version of the proverbial coffeehouse napkins on which so many important inventions reportedly have been born. In a seamless environment, the ideas of conversation can become not only external and directly manipulable as in the Colab, but also portable.

SEAMLESS TECHNOLOGY AND THE RESPONSIVENESS OF ORGANIZATIONS

In this section we want to step back from our general development to compare our thesis with a contrasting vision of computer support for meetings, and also to offer some thoughts on the effects that this kind of technology might have on organizations.

We began our discussion of computer support for meetings with a discussion of the Colab meeting room and the various assumptions that led to its design. The Colab is not the first example of computer support for meetings. Doug Engelbart's early demonstrations (Engelbart, 1984b) during the 1960s and 1970s have become almost legendary. One area of earlier and continuing activity reviewed by Kraemer and King (Kraemer & King, 1986) is the use of computer support for a group decision conference. A decision conference is a conference organized by corporate executives facing a major and strategic decision. Often a specialized consulting team is brought in to conduct the meeting. The facilitation team introduces a formal decision method such as multivariate analysis to the executives and runs a conference around the use of the method. Sometimes, computer tools that support the decision method are provided, coupled with large screen televisions or video projectors to make information publicly visible. In the ideal case, the executive team generates its alternatives, identifies its assumptions, and achieves a consensus for a decision after about two days of intense work. At this point, the conference is over, everyone "goes back to work" to implement the decisions, and the model is discarded.

Similar as this concept of "computer support for meetings" may seem to the concepts that we are pursuing, it is actually a study in contrasts. One immediate difference is that decision conferences are aimed at a specific kind of executive decision meeting, usually involving resource allocation. We have focused on small work groups, especially design teams and research groups, rather than executive teams. A decision conference presupposes that there is a "decision" to be made and that a particular formal approach will provide a rational basis for making the decision. During an executive decision conference, much of the work is in determining and supplying parameters for the formal decision model. Our focus has ranged from formal to informal work processes for which an analytical and mathematical model would be inappropriate. Although tools like Cognoter have semiformal methods and other Colab tools have even more formal methods, there is an emphasis on flexible and informal notations especially in the portable meeting scenarios. Finally, the decision conference presupposes that a single meeting occurs, taking somewhere from a few hours to a couple of days. In contrast, we are concerned with a wider range of time intervals: from a few seconds corresponding to conversational acts as meeting events, to an hour or so corresponding to a single session of a meeting, to a few months corresponding to the continuing and portable meetings of a group project.

Seamlessness has several dimensions relevant to computer support for work. Seamlessness for groups means that tools scale gracefully between individual and group work. Seamlessness across locations means that it should be easy to move information workspaces from one setting to another. Seamlessness across tools means that information used in one tool can be moved easily to another. Seamlessness across time means that there are tools that enable one to easily browse information that was developed over time. Seamlessness across media would enable information from one medium (such as speech) to be referenced or copied to another medium (such as text). Seamlessness in general refers to the ability to manage and move information fluidly.

Software and computer systems for groups will not develop if isolated from other computer systems. Indeed, there is a great incentive to be able to exchange data and to be able to use the same software for meetings and for individual work. We believe that software that extends gracefully from individual work to group work and back again will have advantages. We also believe that appropriate architectural approaches will make this a natural direction for software evolution (but that is a subject for another paper).

When the technological advances necessary for making inexpensive, interactive liveboards is ready, we believe that a revolution in computer use by organizations will follow. Liveboards open up the possibility of computer use for purposes much less formal than is now typical. We believe that people should be able to use a computer as casually and simply as they use a whiteboard now, picking up the "chalk" and doodling.

Some people have referred to computers as a kind of information appliance. An appliance like a toaster is readily available and recognizable. It has a single main function which is recognizable from its presentation to a user. For a toaster, the recognizable display consists of the two slots on the top in which one places sliced muffins or toast. Different appliances present different interfaces according to their function and complexity. What is unusual about computers is that they perform multiple functions. It is as if we expected the same appliance to act either as a toaster, a refrigerator, or a microwave.

This brings us to the notion of the easy path from low functionality to high functionality. Starting from the basic whiteboard, one can imagine a gentle gradient toward increasing capability. The learning gradient should be such that a small investment has a large payoff in increased capabilities, quickly drawing one further into learning and using more. At the basic level, one uses recognizable chalk and an eraser in the usual ways. Simple ends need simple means. After learning about writing and erasing one learns to save and recall the contents of the liveboard. After storage and recall comes moving and copying symbols. Incrementally there is a range of functions that can be learned or not, but the entry point should be simple and familiar and the slope of the learning curve should be gentle. Through simple explorations a user should encounter new possibilities.

Putting liveboards in public places and in casual office use may lead to much greater use of computers in organizations. Much of human learning is by apprenticeship and imitation. A typical user of a computer workstation in a corporate organization is the accountant running a spreadsheet program off by himself in his office. A co-worker is unlikely to see him at work and may have little reason to encounter the arcane computer magic that the accountant uses. Learning by casual watching and imitation is both difficult and unlikely because one must go to the accountant's office, and, furthermore, one cannot easily watch both the computer display (where the action is) and the keyboard (where the control is) at the same time. With liveboard software, it is possible to arrange it so the actions at the locus of activity (the chalk) will be much more visible than with the usual computer workstation. If they are more visible, they will probably be much more imitable so that computer skills could spread more rapidly through an organization.

Imitation of methods could occur even for specialized kinds of software. For example, tools like those used in the Colab can make methods and approaches (such as argumentation spreadsheets) more visible. Even in small organizations, some kinds of meetings are repeated so often that it would be worthwhile to create specialized methods and tools to support them. For example, financial planners could use liveboard-based tools for working through what-if scenarios with their clients. Attorneys could use special meeting tools to work through the terms of standard kinds of contracts. Thus, the methods and styles of argument can be rehearsed, shown, imitated, and reviewed in a seamless medium.

Work life in an organization is filled with conversations and to this end, seamlessness may turn out to be a key to the adoption of technology. Seamlessness facilitates imitation. We predict that conversation technology will spread when it is easy to use and is inexpensive.

We conjecture that seamless technology may lead to the more rapid propagation of ideas in organizations, affecting the resources and speed with which organizations can respond to new situations. Ideas that start in offices can spread to the coffee lounge. Similarly, ideas that arise from interactions in a coffee lounge can spread back to offices. Ideas can flow to or from meetings. In either case, an informal medium with memory can make it easier for people to explain ideas to each other and to combine and compare them. Thus, ideas become more portable between both locations and people.

For technology to be accepted, it must be perceived as filling an important need. For example, until a few years ago the common telephone was a form of office technology with a rude user interface. When a telephone rings, it is like someone is banging on your door, insisting that you drop whatever you are doing and give the caller your immediate attention. Nonetheless, ordinary rude telephones satisfied an important need and gained wide use. There is anecdotal evidence that Colab technology satisfies important needs. Faced with meeting in an unequipped office, a frequent Colab user may jokingly ask "how do I point?". Faced with a shortage of space on an ordinary whiteboard, a Colab user may ask "how do I shrink this?" or "how do I move this over?" Ordinary whiteboards are frustrating after experiencing superior media.

In closing, we return to our original theme of unlocking the creative genius inside of us. Genius takes many forms. It is not just the development of new ideas. There can be genius in negotiation, genius in management, genius in creating coherent plans, genius in all of the things that organizations do. Perhaps this genius can be unlocked by new tools for conversation that respect that we do not work in isolation. In this view, the seamless tools for next generation meetings and conversations may shape our next generation organizations.

ACKNOWLEDGMENTS

This work was supported in part by contract N00140-87-C-9288 from the Defense Advanced Research Projects Agency. The chapter grew out of two talks: J. S. Brown's keynote address at the Conference on Computer Supported Cooperative Work at Austin, Texas in December 1986, and M. Stefik's address at the Symposium on Technological Support for Work Group Collaboration at New York University in May 1987.

The ideas in this chapter have evolved over an extended period of time in conversations with many others at Xerox PARC. Many of the ideas came from other members of the Colab project, including Dan Bobrow, Gregg Foster, Herb Jellinek, Stan Lanning, and Deborah Tatar. Special thanks also to Lucy Suchman and Stu Card for many conversations and insights. Thanks to Gregg Foster, Harley Davis, Giuliana Lavendel, Geoff Nunberg, Lee Sproull, Jeff Shrager, Stan Lanning, Deborah Tatar, and Eric Tribble who read earlier drafts of the chapter and offered helpful suggestions. Thanks also to Steve and Joan Osburn of Osburn Design, for contributing the sketches and for their work in designing the physical facilities of the Colab meeting room.

9
A METHOD FOR EVALUATING WORK GROUP PRODUCTIVITY PRODUCTS

Barry D. Floyd
Jon A. Turner
New York University

INTRODUCTION

Most software products on the market today are tailored to support individual workers. One need only consider a typical text editor with commands for inserting, deleting, and moving text to realize that no provisions have been made in these systems for working with others. For example, one cannot easily make comments on draft text, send them to the author, and have them appear in his or her copy adjacent to the portions to which they refer, attributed to the commentator.

Yet, people rarely work in isolation. They interact, share information, apportion tasks among themselves, monitor each other's performance, and communicate. All of this activity occurs in a cultural and social environment whose quality affects the process and outcomes of the work itself. Furthermore, this work activity is influenced by the characteristics of technological tools themselves.

Improvements in computer and communications technologies have made possible a new class of application systems intended to support group work. The objective of these systems is to support meetings, coordination, project management, co-authoring, decision making, and collaboration on a variety of activities. However, the evaluation of these systems presents a unique set of problems.

The purpose of this chapter is to provide guidance to management in the evaluation and selection of group work support systems. We first consider the evaluation problem in general. How does one go about evaluating any application product? The next section provides a definition of group work and describes the ways group work differs from individual work. We then extend the evaluation approach presented earlier to encompass systems supporting group work. Finally, we demonstrate the use of this evaluation methodology with one hypothetical group application, co-authoring. An application of the methodology to four group support products demonstrated at the Symposium is contained in the Appendix section of this chapter.

THE EVALUATION PROBLEM

Evaluating software for support of group work has two components. The first component involves the normal criteria and methods applied to the evaluation of any application software product. The second component addresses those issues that are unique to group work. Normal criteria applied to application software include such factors as:

Functionality. What features does the product have and how well do they match the work task to be performed?

Integration. How well are the components of the product related to each other?

Interface. How well has the user interface been designed and how does it match the intended user population in the work setting?

Support. How much training and technical assistance will be needed by the intended user population?

Reliability. How well is the product coded, what type of developer support is likely to be needed, will the product be enhanced, and will the developer firm survive?

Efficiency. What resources are needed to run the product and how well does the product fit in with other products in the user's immediate operating environment?

Ease of Modification. What can be changed and how easy is it to do so?

Economics. What are the costs of using the product and how do they compare to the anticipated benefits?

The first three categories are considered functional performance because they identify a product's functionality and the methods by which this functionality is invoked (i.e., attributes that are specific to the product). Other criteria, such as reliability and efficiency, are generic to all products.

Much has been written about these subjects and it is not our intent to repeat or summarize that material here. It is sufficient to observe that:

1. The intended use of a product and the specific operating environment determine which of these factors predominate in an evaluation.

2. The above list of criteria is not closed. Other categories could easily be added with proper arguments as to their importance.

3. Although application software evaluation does have objective aspects, the process is essentially subjective. That is, the importance assigned to any factor is a function of individual assumptions and opinions that are often implicit and not divulged by the analyst.

4. The tendency is to perform an incremental evaluation; that is, compare one product to another in a product category, rather than using a top down approach where needs are determined and then compared with the available products. Although incremental evaluation may result in a local optimum, it often provides a major error.

The evaluation of applications to support group work will be even more difficult than selecting systems for individuals. This is because group work extends the needed features and coverage of a system in new directions, creating a different, larger evaluation problem. Attention is now turned to the characteristics of group work.

NATURE OF GROUP WORK

Group work differs from individual work in four respects:

1. It involves extensive and varied person-to-person communication. Consequently, any product supporting group work must have a variety of communications capabilities that can be laid on top of physical services. Furthermore, these capabilities need to be customized for individual workers. In addition, some entity has to manage the communications activity.

2. There are processes, beyond those necessary for individual workers, that need to be supported. Roles, protocols, and procedures have to be established. Group process needs to be monitored and facilitated. Interpersonal activities must be attended to.

3. Because the overall work task is partitioned among a number of workers, there is a task management function that must be performed. This involves assignment, status monitoring and integration of individual work products, and their assembly into a deliverable unit.

4. The relationship of the work group to the larger entity, of which it is part, needs to be considered. This includes such issues as organizational culture, structure, power, authority, norms and values.

There are many groups within organizations. Examples are:

- Committees
- Project teams
- Formal authority groups

- Peer groups
- Informal social groups or networks
- Information exchange networks

The *project team* is used for the purposes of discussion here because it is frequently found in business organizations and has both formal and informal components. A team can be defined as:

> ... a group of individuals (two or more) where the members assume specialized roles in both doing the work and maintaining the cohesiveness and morale of the team members. We assume the team has the resources needed to fulfill its role and functions, that the team has the ability to use these resources at the appropriate times and levels, and that they have the capacity to plan, organize, make decisions, communicate and negotiate the actions to reach the team's objectives. (Cashman, 1987, p.10)

Teams are characterized by much communication for the purpose of information sharing, resource allocation, problem solving, and negotiation over courses of action (Semprevino, 1980). Individuals working alone are spared this communication load because most of these activities are performed by themselves and whatever communication is needed takes place internally, presumably at much higher bandwidth and with much greater understanding (e.g., in some efficient internal representation).

EVALUATION METHODOLOGY FOR GROUP WORK PRODUCTS

Because of the need to support the individual's enhanced role as a member of a group, normal evaluation methods must be extended to encompass group support application systems. Two levels of evaluation are considered: coarse and detailed.

Coarse Evaluation

Coarse evaluation refers to the coverage of a system, both in features provided by the application software, and in connectivity or topology of the work group supported. Application systems must exceed a critical mass of coverage for them to be interesting. If the software contains insufficient features, or does not cover a major part of the work group, and often key individuals outside of the immediate group, it is not worthwhile to do a detailed analysis (See Figure 9.1).

Broad feature coverage can be determined by comparing the general features of the application system to a classification scheme, such as that provided by Johansen and described in chapter 1 of this book (see Chapter 1, Table 2). Connectivity

Categories	System			
---	A	B	C	D
Face to Face Meeting Support				
Facilitation services				
Group decision support	X			
Presentation support	X			
Computer supported meetings				
Support for Electronic Meetings				
Extensions of telephone		X		
Personal computer software				
Computer conferencing		X		X
Text filtering				
Audio or video teleconferencing				X
On-line resources				
Support Between Meetings				
Project management				
Calendar management				
Group writing software				
Conversational structuring				
Text filtering				
Spontaneous interaction				
Comprehensive support			X	

Figure 9.1. Group support system coverage diagram

coverage can be determined by comparing those who will use the system with the composition of the work group.

Detailed Evaluation

Although the focus of this evaluation method is on group work, at the most primitive level most work is performed by individuals. Consequently, a major portion of the evaluation process for a group support application product must consider how well that product supports an individual performing his/her work activity, whatever that may be.[1] The domain over which this evaluation takes place

---

1   Subtle re-structuring of individual work activities may take place when a new, computer-based tool is used in performing the task. For example, if the job were assembling a box and the tool was changed from a hammer to a screwdriver, the task sequence and activities would change.

is called *task*. But this is not sufficient for the evaluation of group support systems. Two new domains must also be considered: group processes and communications.

Group processes are those activities that support and facilitate individuals working together. For a team this might consist of agreeing on a leader; working out key group processes, such as the way of deciding important questions, or setting work agendas; and establishing a method of resolving conflicts when they arise. It is in providing an individual support for his or her role in a group that this technology has its greatest potential.

Communications are those exchanges between two or more people that result in transfers of information. These consist of the ability to send and receive messages, methods of addressing individuals, network topologies, characteristics, and protocols.

Not only must functional performance of the system be evaluated, but two additional categories need to be considered: the extent to which the system supports administrative activities and how well it fits the organization. Figure 9.2 shows the detailed evaluation matrix.

Functional Performance refers to how well the application system functions in each domain. For example, functional performance in the task domain refers to how well the system supports individual work tasks. This is similar to evaluating a normal application system for a single user where one is concerned with the features supported, the degree of integration of these features, and the quality of the interface. Performance in the group process domain refers to how well the system supports those activities that are unique to group work, for example, group decision making, or the sharing of a work product for mutual comment. Performance in the communications domain refers to the features provided by the application to support communications, for example, whether a return receipt can be requested and whether one can reply to a message.

Administration refers to how well the application system supports the management of each domain. In other words, how well does the system keep track of resources available, assigned, and used; how easy is it to enter this information; and how is status determined.

Organizational Fit refers to how well an application system compares with the explicit or implicit procedures, policies, norms, and values of the task, communication, and group process domains of the organization for which it is intended. Any designed object, including application products, have a central concept or theme. Philosophies of designers shape this central concept through decisions that produce features and procedures of the application product. The result is an implicit structuring of tasks through the specification of which can be performed by whom, when. Organization designers, as well as organization culture, influence policy and procedure decisions that establish norms of accepted behavior. These organizational norms interact with those of individuals to produce work group

		Domain	
Categories	Task	Group	Communication
functional performance	Normal System		
admin			
org fit			

Figure 9.2: Detailed evaluation matrix for group support systems

policies, procedures and norms. Fit refers to how well the system supports existing individual and organizational norms and procedures.

Organizational fit consists of at least four components: philosophy, work process, structure, and control. Philosophy is the dominant style of an organization often expressed by the behavior of its leaders, for example, authoritarian, democratic, laissez faire or paternalistic. *Work process* is the sequence in which tasks are performed and their organization. *Structure* refers to relationships among workers. *Control* refers to strategies used to control worker behavior and performance.

An example may clarify the notion of organizational fit. Suppose one were evaluating a presentation support system for a military command unit. The application system has a democratic design concept that lets a person message anyone else with their comments about the presentation. The functional performance and administrative support capabilities of the system are considered adequate. However, the democratic concept of the system, and specifically the ability to generate and transmit comments independent of role or level of authority of the commentator, conflicts with the military's notion of chain of command that routes information along specified channels based on role and where decision authority and responsibility rest with one individual. Thus, there is a lack of fit in philosophy (system is democratic whereas the organization is authoritarian and the individual is not considered) and in structure (system permits lateral relationships whereas the organization does not) while in control there is more of a fit (because the system keeps a formal record of what were previously informal communications facilitating assignment of responsibility).

EVALUATION EXAMPLE

A hypothetical example is used to illustrate the proposed evaluation framework.

Categories	System Co-Author
Face to Face Meeting Support	
Facilitation services	
Group decision support	
Presentation support	
Computer supported meetings	
Support for Electronic Meetings	
Extensions of telephone	
Personal computer software	
Computer conferencing	X
Text filtering	
Audio or video teleconferencing	
On-line resources	
Support Between Meetings	
Project management	X
Calendar management	X
Group writing software	X
Conversational structuring	
Text filtering	
Spontaneous interaction	
Comprehensive support	

Figure 9.3. Co-Author coverage diagram

Co-Author

Co-Author is a hypothetical system that supports two or more authors working on the same paper. It permits multiple reviewers to comment on portions of text and to have these comments be visible to readers of the paper attributed to reviewers. The system also allows both concurrent and non-concurrent conferencing to discuss the paper and a reviewer's comments with one or more of the authors. The system keeps track of various versions of the paper as well as proposed revisions. It contains a formatter and interfaces directly with a high quality output device as well as a PBX. In this system, a paper goes through several stages, among them "draft" and "final." Changes can only be made in the draft stage and only by the person who originally entered the data. In addition, an agenda must be created prior to establishing a computer conference.

Figure 9.3 provides the coverage diagram for Co-Author. The system provides facilities for supporting group management and project management in the sense

Domain

Categories	Task	Group	Comm
functional perf	editing features formating features	agenda setting feedback to reviewers exchanges among co-authors	send/receive copy co-authors and reviewers
admin	assing ind work tasks status of ind tasks	aportion work among co-auth status of co-auth relations control over revisions	conn to co-auth, rev status of msg loc of players conf
org fit	work proc "draft" "final" change only draft	only orig auth change draft agenda prior to conf	auth and rev interconn

Figure 9.4: Detailed evaluation matrix for Co-Author

of a paper being a "project." It also provides computer conferencing and some calendar management for conference scheduling.

Figure 9.4 shows the detailed coverage diagram for Co-Author. At the functional performance level, the system contains normal editing and formatting features to support individual authoring. In addition, it supports group activities such as agenda setting among the co-authors and the establishment of agendas when an electronic conference is planned. It also provides facilities for commenting back to reviewers and for exchanges among co-authors. In the communications domain, the system supports message and file transfer between authors and reviewers. At the administrative level, the system supports the creation of individual work tasks, entering status data about them and reporting project status. It also provides facilities for assigning work among authors and to reviewers, determining the status of the paper, author or reviewers, and control over which revisions are entered into the draft and final versions of the paper. In the communications domain, the system maintains track of the locations of authors and reviewers, the status of outstanding messages and the history of messages between partners.

At the organizational fit level, there are no serious philosophy conflicts, but a rigid work process that does not permit changes at the final phase (task domain) and strict control that only allows the author of a section to enter revisions (group domain) makes the application unacceptable for the work group that is considering using the product.

Four examples applying this evaluation methodology to current group support applications are provided in the Appendix.

CONCLUSION

It has been argued here that evaluating work group support products is a special case of evaluating any computer software. That is, a number of factors must be considered over a number of domains. Group work has been shown to differ from individual work in a variety of ways, particularly in the making of many internal processes explicit. The notion of functional performance across the task domain was expanded to include two new domains of group and communications. Two new categories, administration and fit, were also introduced as levels or classes of evaluation. The methodology was applied to a hypothetical example of collaborative authoring.

Although this methodology is by no means complete, it is an attempt to evaluate application products on those dimensions that differentiate group from individual work.

APPENDIX: PRODUCT EVALUATION

Metaphor

METAPHOR is a comprehensive system that includes equipment, software and communications offering users a set of integrated tools designed to support data retrieval, analysis, and text processing. The system consists of high resolution graphics workstations interconnected via a local area network (LAN). This configuration allows for local storage at the workstation and shared data storage via a file server on the LAN. Users may communicate through a messaging facility over the network. Messages can be comprised of text, programs, models, and/or graphics.

Figure 9.5 provides the coverage diagram for METAPHOR. The system's strength is in supporting an individual (typically a brand manager). This support is directed toward activities that take place between meetings; the system offers little direct support for real-time meetings, electronic or face-to-face.

Figure 9.6 shows the detailed evaluation matrix for METAPHOR. At the functional performance level, METAPHOR mainly supports data-retrieval and analysis tasks. It augments these capabilities by allowing a user to combine them into larger, executable modules. For example, a user can define a data retrieval task, have the results sent to a spreadsheet model, and have these results incorporated into a report. This sequence can be defined, saved, and then re-executed at any time. Further, it may be sent to another user over the network.

An important aspect of the system is its ease of use. The interface is icon-based. The design goal was to allow non-computer professionals to perform their own in-

Categories	System			
	M	IS	IB	C
Support for face-to-face meetings				
Facilitation services				
Group decision support				
Presentation support software		X		
Computer-supported meetings				
Support for electronic meetings ·				
Extensions of the telephone				
Personal computer software				
Computer conferencing		X	X	
Text filtering				
Assistance for teleconferencing				
On-line resources				
Support between meetings				
Project management software				
Calendar management software				X
Group writing software				
Conversational structuring				X
Group memory management				
Spontaneous online interaction				
Comprehensive support systems	X			

```
M  - METAPHOR
IS - INSYNCH
IB - InBOX
C  - COODINATOR
```

Figure 9.5: Group support system coverage diagram

formation-processing tasks quickly and easily through direct manipulation of objects. It is highly successful at this and provides an almost ideal work environment.

At the administrative level, METAPHOR provides little support. There are no project or individual work management functions. For example, version control of models, memos, and the like must be maintained manually by the individual user, as must communication and distribution information.

At the organizational fit level, METAPHOR does not provide any explicit structuring of tasks or communications processes. It is likely the LAN will interconnect those workers who are in proximity to one another and therefore have relationships already established; it does not really support an extended community.

An underlying philosophy in METAPHOR is that a user will be performing his or her own work on the system without the need of a data-processing intermediary. The situation where the system would thrive is one where an individual

Categories	Domain		
	Task	Group	Comm
functional perf	text processing graphics data retrieval data analysis	allows sharing	local area network messaging system
admin	-	-	-
org fit	individual performs his own tasks	no struct of work process	physically close no struct of com process

Figure 9.6: METAPHOR detailed evaluation matrix

user has extensive data analysis needs, an inclination to use a system personally, and a constraint in using information services. Although there is no explicit need to obtain a critical mass of users greater than one, more users are needed to take advantage of information sharing.

Insynch

INSYNCH is a real-time teleconferencing software package for microcomputer users connected via standard telephone lines. Users on each end of the line may view and interact with the same application, such as a sales forecast developed in the LOTUS 123 spreadsheet package. To provide this teleconferencing capability, the software captures keystrokes from the keyboards of each user and sends them to both processors. If both computers have the same software and data files, then the same operations occur on each system (provided synchronization is not lost).

Figure 9.5 shows the coverage diagram for INSYNCH. The system supports electronic meetings and is typical of screen sharing programs.

Figure 9.7 shows the detailed evaluation matrix for INSYNCH. At the functional performance level, INSYNCH offers a number of interesting features in teleconferencing. The first is a screen presentation subsystem where users may capture, annotate, and save any screen displayed on their computer. These screens may be organized into a structured presentation along with screens developed with a screen generation package. A second feature is file transfer between microcomputers. This allows users to assure that the data being used is the same on each com-

		Domain	
Categories	Task	Group	Comm
==============	===================	===============================	===================================
functional perf	presentation prep	presentations screen sharing	2 person comm link
admin	-	minutes	file transfers to share data/programs
org fit	-	little control mechanisms	2 persons only

Figure 9.7: INSYNCH detailed evaluation matrix

puter. Furthermore, INSYNCH is easy to use, providing a menu-based interface that is simple to understand and easy to manipulate.

At an administrative level, INSYNCH provides a "minutes" or log function, recording detailed records of transactions occurring during a teleconference.

At the organizational fit level, INSYNCH provides new communication channels for two users who are geographically separated. The meeting must take place in real time. INSYNCH does not explicitly structure tasks or communications. Although the system does provide the infrastructure for display and interaction of two users working on the same task, there are few mechanisms to control communication. Much like the telephone, two users may speak at the same time. And though the impact of inappropriate timing may result in a perception of rudeness in telephone conversation, in real-time computing environments with keystrokes being entered, it results in nonsense instructions being sent to the system. In viewing and revising a spreadsheet, this lack of control becomes potentially dangerous to unprotected cells. If problem solving and other interactive types of conversation are to occur with its concomitant interruptions, simultaneous beginnings, and other confusions, additional mechanisms for control and synchronization (such as easily going back to point x) need to be included in the software.

InBOX

InBOX is a MacIntosh-based messaging system that allows users to send and receive messages over a local area network. InBOX is easy to learn, with an icon-based interface; the goal is to have an unskilled person able to use the system within 10 to 15 minutes. The expected set of users includes both professionals and secretaries. This differing level of computer expertise and skill led to a system interface designed to be comfortable for both. The primary concern was to have a

| | Domain | | |
Categories	Task	Group	Comm
functional perf	easy to use	-	messaging
admin	-	-	mail lists
org fit	-	-	close proximity due to LAN

Figure 9.8: InBOX detailed evaluation matrix

critical mass of users, so that most people would find others they needed to communicate with on the system. Figure 9.5 shows the coverage diagram for InBOX.

From a functional performance perspective, the package is similar to other messaging systems. Its market niche is MacIntosh users. The system provides relatively unstructured formats except for phone and memo messages. The detailed evaluation matrix for InBOX is shown in Figure 9.8.

InBOX is consistent with the Macintosh style of user interaction. Skills learned in this environment carry over to this package, reducing the amount of training required.

From an administrative perspective, InBOX offers mail lists to facilitate the sending of messages to groups of people and an RSVP function that alerts the sender when a message has been opened by the receiver.

No special organizational demands are made of the system; it allows communications among any members of the network. It also runs on a local area network (one of the processors acts as the message system controller) that limits its scope within the organization to those individuals connected to the network.

The COODINATOR

The COORDINATOR is a communications tool that structures conversations people have with each other. The designers of this product assert that conversations are structured (or should be) and that when this structure is explicit the process of communication is improved.

Conversations take place within domains. A domain is a central organizating concept or topic (e.g., hiring personnel) for a particular set of conversations. Two kinds of conversations within a domain are identified: those for action and those for possibilities. Conversations for action are those in which participants' inten-

Domain

Categories	Task	Group	Comm
functional perf	writing tools	-	sending/receiving messages
	calendar mgmt		
admin	-	-	manages names on network organizes messages as conversations
org fit	-	-	structures messages and conversations

Figure 9.9: The COORDINATOR detailed coverage diagram

tions are to produce future actions. Conversations for possibilities are ones in which the result is not a commitment for action, but the possibility for one.

Within these conversations, four things can occur: requests, promises, assertions, and declarations. In conversations for action, the primary occurrences are requests and promises (e.g., You request an expansion plan and I promise to give it to you next Tuesday). In conversations for possibilities, the primary occurrence is a declaration (e.g., It is possible that John and George could work together on this project). The COORDINATOR uses this framework to structure conversations.

Figure 9.5 provides the coverage diagram for The COORDINATOR. The system is a conversation structuring tool supporting activities between meetings. The COORDINATOR runs on personal computers connected via a communications network such as a telephone line or a local area network; one dedicated system serves as the "hub" or message server.

Figure 9.9 shows the detailed evaluation matrix for The COORDINATOR. At the functional performance level, The COORDINATOR provides writing tools and calendar management for individual tasks. For communications it offers sending and receiving of messages to others with access to the network and running COORDINATOR software. At the administrative level, in the task domain, it provides support for storing and retrieving documents and calendar information. For communications, it has a directory containing the names and network addresses of people with whom a user has had or is likely to have conversations. Conversations are linked so that they can be recalled and reviewed together. At the organizational fit level the most important and interesting aspect of The COORDINATOR is the structure it imposes on conversations. The system requires a particular type of response to a request (e.g., a commitment) and specific information that is associated with that response (e.g., date when the commitment will be completed).

This structure removes ambiguity which occurs in many conversations and makes commitments explicit. It shows clearly when commitments are not being kept and makes it more difficult to "let things slide." If the intention is to commit without really performing, then this becomes clear. However, not all participants may be willing to make their commitments this explicit. And the system may promote process at the expense of content. The organizational fit issue is then between this type of conversation structuring and the way group members prefer to interact. There is a certain formality, an unforgiving nature, to The COORDINATOR; it does not permit the kind of slack and flexibility that usually occurs and is realistically needed in day-to-day work.

PRODUCT EVALUATION SUMMARY

The four packages reviewed in this chapter vary with respect to their support for group work. Three are explicitly communications-oriented. Two of these have the establishment of new communications channels as their main objective (InBOX and INSYNCH) and these facilitate group interaction through enhanced functionality in data sharing and messaging. Neither offers explicit support of group process.

METAPHOR's focus is on the individual in providing an integrated and user-friendly analysis environment. It impacts the group only by shifting information processing power to non-IS professionals. Perhaps the most interesting system from the perspective of this chapter is The COORDINATOR. It attempts to redefine the components of communication and to make them explicit; thus it provides not only functionality, but a whole communications philosophy.

REFERENCES

Akin, G., & Hopelain, D. (1986). Finding the culture of productivity. *Organizational Dynamics, 14(3)*, 19-32.

Anshen, R. N. (1973). Introduction. In I. Illich (Ed.), *World perspectives series, tools for conviviality* (Vol. VII), New York: Harper & Row.

Arnold, C. (1968, Fall). Oral rhetoric, rhetoric, and literature. *Philosophy and rhetoric*, 191-210.

Arrow, K. (1974). *Limits of organization*. New York: Norton.

Beer, S. (1972). *The brain of the firm*. New York: Herder & Herder.

Begeman, M., Cook, P., Ellis, C., Graf, M., Rein, G., & Smith, T. (1986). Project NICK: Meeting augumentation and analysis. *Proceedings of the Conference on Computer-Supported Cooperative Work* (1-6). Austin, TX: Microelectronics and Computer Technology Corp.

Bernstein, P., & Goodman, N. (1981). Concurrency control in distributed database systems. *ACM Computing Surveys, 13*, 185-232.

Bikson, T. K. (1987). Understanding the implementation of office technology. In Kraut, R. (Ed.), *Technology and the transformation of white collar work* (155-176). Hillsdale NJ: Lawrence Erlbaum Associates.

Bikson, T. K., & Eveland, J. D. (1986). *New office technology: planning for people*. New York: Pergamon Press.

Bikson, T. K., & Gutek, B. A. (1983). *Advanced office systems: An empirical look at utilization and satisfaction* (Technical Report N-1970-NSF). Santa Monica, CA: The RAND Corporation.

Bikson, T. K., & Gutek, B. A. (1984). Training in automated offices: An empirical study of design methods. In Rijnsdorp, J. W. & Plomp, T. J. (Eds.), *Training for tomorrow* (129-145). New York: Pergamon Press.

183

Bikson, T. K., Gutek, B. A., & Mankin, D. A. (1981) *Implementation of information technology in office settings: Review of relevant literature* (Technical Report P-6697). Santa Monica, CA: The RAND Corporation.

Bikson, T. K., Gutek, B. A., & Mankin, D. A. (1987). *Implementing computerized procedures in office settings: Influences and outcomes* (Technical Report R-3077-NSF). Santa Monica, CA: The RAND Corporation.

Bikson, T. K., Stasz, C., & Mankin, D. A. (1985). *Computer-mediated work: Individual and organizational impact in one corporate headquarters* (Technical Report R-3308-OTA). Santa Monica, CA: The RAND Corporation.

Blomberg, J. (1986). The variable impact of computer technologies on the organization of work activities. *Proceedings of the Conference on Computer-Supported Cooperative Work* (36-42). Austin TX: Microelectronics and Computer Technology Corp.

Cashman, M. (1987). *Improving team productivity and effectiveness* (Technical Report). Morristown, NJ: Cashman Consulting Corporation.

Cashman, P., & Stroll, D. (1987). Achieving sustainable management of complexity: A new view of executive support. *Office: Technology and People, 3*, 147-173.

Chandler, A. (1962). *Strategy and structure*. Cambridge, MA: MIT Press.

Conklin, J. (1987). Hypertext: An introduction and survey. *IEEE Computer, 20*, 17-41.

Croft, W. B., & Lefkowitz, L. (1984). Task support in an office system. *ACM Transactions on Office Information Systems, 2*, 115-141.

Crowston, K., Malone, T., & Lin, F. (1986). Cognitive science and organizational design: A case study of computer conferencing. *Proceedings of the Conference on Computer-Supported Cooperative Work* (43-61). Austin TX: Microelectronics and Computer Technology Corp.

Dalton, R. (1987, March 9). Group-writing tools: Four that connect. *Information Week*, 62-65.

De Cindio, F., De Michelis, G., Simone, C., Vassallo, R., & Zanaboni, A. (1986). CHAOS as a coordination technology. *Proceedings of the Conference on Computer-Supported Cooperative Work*, (325-342). Austin, TX: Microelectronics and Computer Technology Corp.

DeJong, G. F. (1979). Prediction and substantiation: A new approach to natural language processing. *Cognitive Science, 3*, 251-273.

DeLong, D., & Rockart, J. (1984). *A survey of current trends in the use of executive support systems* (Working paper 121). Cambridge, MA: Center for Information Systems Research, Massachusetts Institute of Technology.

Denning, P. (1982). Electronic junk. *Communications of the ACM, 23(3)*, 163-165.

DeSanctis, G. & Gallupe, R. (1987). A foundation for the study of group decision support systems. *Management Science, 33(5)*, 589-609.

Dhar, V. & Pople, H. (1987). Rule-based versus structure-based models for explaining and generating expert behavior, *Communications of the ACM, 30(6)*, 542-555.

Dhar, V., & Ranganathan, P. (1986, October). Automating review of forms for international trade transactions: A natural language processing approach. *Proceedings of the Third International Conference on Office Information Systems* (61-69). New York: Association for Computing Machinery.

Diehl, M. & Stroebe, W. (1987). Productivity loss in brainstorming groups: Toward the solution of a riddle, *Journal of Personality and Social Psychology, 53(3)*, 497-509.

Dunham, R., Johnson, B., McGonagill, G., Olson, M., & Weaver, G. (1986). Using a computer based tool to support collaboration: A field experiment. *Proceedings of the Conference on Computer-Supported Cooperative Work* (343-352). Austin TX: Microelectronics and Computer Technology Corp.

El Sawy, O. (1985). Personal information systems for strategic scanning in turbulent environments: Can the CEO go on-line? *MIS Quarterly, 9(1)*, 53-59.

Ellis, C. (1979). Information control nets: A mathematical model of office information flow. *Proceedings of the ACM Conference on Simulation, Modeling, and Measurement of Computer Systems* (225-240). New York: Association for Computing Machinery.

Emery, F.E., & Trist, E.L. (1973). *Towards a social ecology*. London: Plenum Press.

Engelbart, D. (1963). A conceptual framework for the augmentation of man's intellect. In Howerton, P.W. & Weeks, D.C. (Eds.), *Vistas in information handling* (1-29). Washington DC: Spartan Books.

Engelbart, D. (1984). Collaboration provisions in Augment. *Proceedings of the AFIPS Office Automation Conference* (51-58). Washington, DC: AFIPS Press.

Englebart, D. C., & English, W. K. (1968). Research center for augmenting human intellect. *Proceedings of Fall Joint Computing Congress* (395-410). Washington, DC: AFIPS Press.

Engelbart, D., Watson, R., & Norton, J. (1973). The augmented knowledge workshop. *Proceedings of the AFIPS National Conference* (9-21). Washington, DC: AFIPS Press.

Eveland, J. D., & Bikson, T. K. (1987). Evolving electronic communication networks: An empirical assessment. *Office: Technology and People, 3*, 103-128.

Eveland, J. D., & Bikson, T. K. (1988) Work group structures and computer support: A field experiment (Technical Report WD-3974-MF). Santa Monica, CA: The RAND Corporation.

Feldman, M. (1987). Electronic mail and weak ties in organizations. *Office: Technology and People, 3*, 83-102.

Fikes, R. (1982). A commitment-based framework for describing cooperative work. *Cognitive Science, 6*, 331-347.

Fikes, R., & Kehler, T. (1985). The role of frame-based representation in reasoning. *Communications of the ACM, 28(7),* 904.

Flores, F., & Bell, C. (1984, Fall). A new understanding of managerial work improves system design. *Computer Technology Review.*

Flores, F., & Ludlow, J. (1981). Doing and speaking in the office. In Fick, G. & Sprague, R. (Eds.), *DSS: Issues and challenges* (315-337). London: Pergamon Press.

Fox, M. (1981). An organizational view of distributed systems. *IEEE Transactions on Systems, Man, and Cybernetics, SMC-11,*1, 70-80.

Galbraith, J. (1973). *Designing complex organizations.* Reading, MA: Addison-Wesley.

Garrett, L., Smith, K., & Meyrowitz, N. (1986). Intermedia: Issues, strategies, and tactics in the design of a hypermedia document system. *Proceedings of the Conference on Computer-Supported Cooperative Work* (163-174). Austin, TX: Microelectronics and Computer Technology Corp.

Gatlin, L.L. (1972). *Information theory and the living system.* New York: Columbia University Press.

Ghoshal, S., & Kim S. (1986). Building effective intelligence systems for competitive advantage. *Sloan Management Review, 28,* 49-58.

Gifford, D. K., Baldwin, R. W., Berlin, S. T., & Lucassen, J. T. (1985, December). An architecture for large scale information systems. *Proceedings of the 10th ACM Symposium on Operating Systems Principles* (161-170). New York: Association for Computing Machinery..

Goodman, G., & Abel, M. (1986). Collaboration research in SCL. *Proceedings of the Conference on Computer-Supported Cooperative Work* (246-252). Austin, TX: Microelectronics and Computer Technology Corp.

Gray, P. (1986). Group decision support systems. In McLean, E. R., & Sol, H.G. (Eds.), *Decision support systems: A decade in perspective.* Amsterdam: North-Holland.

Greif, I. (1982). *Cooperative office work, teleconferencing, and calendar management: A collection of papers (unpublished technical memo).* Cambridge, MA: Laboratory for Computer Science, Massachusetts Institute of Technology.

Greif, I., & Sarin, S. (1986). Data sharing in group work. *Proceedings of the Conference on Computer-Supported Cooperative Work* (175-183). Austin, TX: Microelectronics and Computer Technology Corp.

Gutek, B. A., Bikson, T. K., & Mankin, D. A. (1984). Individual and organizational consequences of computer-based office information technology. *Applied Social Psychology Annual, 5,* 231-254.

Gutek, B. A., Sasse, S. H., & Bikson, T. K. (1986). The fit between technology and workgroup structure: The structural contingency approach and office automation. *Proceedings of the Academy of Management.* New York: National Academy of Management.

Halasz, F., Moran, T., & Trigg, R. (1987). Notecards in a nutshell. *Proceedings of the ACM Conference on Computer Human Interaction* (45-52). New York: Association for Computing Machinery.

Hiltz, S. R., & Turoff, M. (1978). *The network nation: human communication via computer.* Reading, MA: Addison-Wesley.

Hiltz, S. R., & Turoff, M. (1985). Structuring computer-mediated communication systems to avoid information overload. Communications of the ACM, 28(7), 680-689.

Holt, A. (1986). *Coordination technology and Petri nets.* (Technical report). Trumbull, CT: Coordination Technology, Inc.

Holt, A., & Cashman, P. (1981). Designing systems to support cooperative activity: An example from software maintenance management. *Proceedings of the Fifth Computer Software and Applications Conference* (192-203). Los Alamitos, CA: Institute of Electrical and Electronics Engineers.

Holt, A., & Ramsey, H. R. (1985). *Coordination systems: The User's view (Technical report).* Trumbull, CT: Coordination Technology, Inc.

Huber, G. (1984). The nature and design of post-industrial organizations. *Management Science, 30,* 928-951.

Huber, G., & McDaniel, R. (1986). The decision-making paradigm of organizational design. *Management Science, 32,* 72-589.

Humphreys, P. (1984). Levels of representation in structuring decision problems. *Journal of Applied Systems Analysis, 11,* 3-21.

Humphreys, P., & Berkeley, D. (1983). Problem structuring calculi and levels of knowledge representation in decision making. In Scholz, R. (Ed.), *Decision making under uncertainty* (121-157). Amsterdam: Elsevier Science Publishers B. V.

Johansen, R., Vallee, J., & Vian, K. (1979). *Electronic meetings.* Reading MA: Addison Wesley.

Johansen, R. (1984). *Teleconferencing and beyond: Communications in the office of the future.* New York: McGraw-Hill.

Johnson, B. (1981). *Communication: The process of organizing.* Boston: American Press.

Johnson, B. (1986). *Growing productive information systems: The case for participative learning environments.* Paper presented to Work in America Institute/The Productivity Forum, Scarsdale NY.

Johnson, B.M. & Rice, R. (1987). *Managing organizational innovation: The evolution from word processing to office information systems.* New York: Columbia University Press.

Kedzierski, B. (1983). *Knowledge-based communication and management support in a system development environment (Technical report KES.U.83.3).* Palo Alto, CA: Kestrel Institute.

Keen, P., & Scott Morton, M. (1978). *Decision support systems: An organizational perspective.* Reading, MA: Addison Wesley.

Kling, R., & Scacchi, W. (1982). The web of computing: Computer technology as social organization. *Advances in Computers, 21,* 2-60.

Kraemer, K., & King, J. (1986). Computer-based systems for group decision support: Status of use and problems in development. *Proceedings of the Conference on Computer-Supported Cooperative Work* (353-375). Austin, TX: Microelectronics and Computer Technology Corp.

Kraut, R., Galegher, J., & Egido, C. (1986). Relationships and tasks in scientific research collaborations. *Proceedings of the Conference on Computer-Supported Cooperative Work* (229-245). Austin TX: Microelectronics and Computer Technology Corp.

Lowe, D. (1986). SYNVIEW: The design of a system for cooperative structuring of information. *Proceedings of the Conference on Computer-Supported Cooperative Work* (376-386). Austin, TX: Microelectronics and Computer Technology Corp.

Luconi, F., Malone, T., & Scott Morton, M. (1985). *Expert systems and expert support systems: The next challenge for management (Technical report 85-005).* Cambridge, MA: Management in the 1990s Program, Massachusetts Institute of Technology.

Malone, T. (1985). Designing organizational interfaces. *Proceedings of the ACM Conference on Computer-Human Interaction (66-72). New York: Association for Computing Machinery.*

Malone, T., & Smith, S. (1984). T*radeoffs in designing organizations: Implications for new forms of human organizations and computer systems (Technical report 112).* Cambridge, MA: Center for Information Systems Research, Massachusetts Institute of Technology.

Malone, T., Yates, J., & Benjamin, R. (1987). Electronic markets and electronic hierarchies. *Communications of the ACM, 30,* 484-497.

Malone, T., Grant, K., Lai, K., Rao, R., & Rosenblitt, D. (1987). Semi-structured messages are surprisingly useful for computer-supported coordination. *ACM Transactions on Office Information Systems, 5,* 115-131.

Malone, T.W., Grant, K.R., Turbak, F.A., Brobst, S.A., & Cohen, M.D. (1987). Intelligent information sharing systems, *Communications of the ACM, 30,* 390-402.

Marca, D., & Cashman, P. (1985). Toward specifying procedural aspects of cooperative work. *Proceedings of the Third International Workshop on Software Specification* (151-154). Los Alamitos, CA: Institute of Electrical and Electronics Engineers.

Marca, D., Schwartz, S., & Casaday, G. (1987). A specification method for cooperative work. *Proceedings of the Fourth International Workshop on*

Software Specification (242-248). Los Alamitos, CA: Institute of Electrical and Electronics Engineers.

March, J. G. & Simon, H. A. (1958). *Organizations.* New York: John Wiley & Sons.

McCune, B. P., Tong, R. M., Dean, J. S., & Shapiro, D. G. (1985). RUBRIC: A system for rule-based information retrieval. *IEEE Transactions on Software Engineering, 11,* 939-944.

McGrath, J. E. (1984). *Groups: Interaction and performance.* Englewood Cliffs NJ: Prentice-Hall.

McGrath, J. E., & Altman, I. (1966). *Small group research.* New York: Holt, Rinehart and Winston.

Meade, J.E. (1971). *The controlled economy.* London: George Allen and Unwin Ltd.

Meador, C., Guyote, M., & Rosenfeld, W. (1986). Decision support planning and analysis: The problems of getting large-scale DSS started. *MIS Quarterly, 10,* 159-177.

Miles, R.E., & Snow, C.C. (1986). Networked organizations: New concepts for new forms. *The McKinsey Quarterly, 3,* 53-66.

Mintzberg, H. (1979). *The structuring of organizations.* Englewood Cliffs, NJ: Prentice-Hall.

Natanson, M. (1970). *The journeying self: A study in philosophy and social role.* Reading, MA: Addison-Wesley.

Neches, R. (1986). Tools help people cooperate only to the extent that they can help them share goals and terminology. *Proceedings of the Conference on Computer-Supported Cooperative Work* (192-202). Austin, TX: Microelectronics and Computer Technology Corp.

Nelson, T. (1981). *Literary machines.* Swarthmore PA: available from author.

Nilles, J., El Sawy, O., Mohrman, A., Jr., & Pauchant, T. (1986). *The strategic impact of information technology on managerial work (Technical report).* Los Angeles: Center for Futures Research, Graduate School of Business Administration, University of Southern California.

Noreault, M. K., & McGill, M. J. (1977). Automatic ranked output from boolean searches in SIRE. *Journal of the ASIS, 28,* 333-339.

Orr, J. (1986). Narratives at work: Story telling as cooperative diagnostic activity. *Proceedings of the Conference on Computer-Supported Cooperative Work* (62-72). Austin TX: Microelectronics and Computer Technology Corp.

Palme, J. (1984). *You have 134 unread mail do you want to read them now?* Paper presented to IFIP Conference on Computer Based Message Services, Nottingham University.

Pava, C. H. (1983). *Managing new office technology.* New York: The Free Press.

Pava, C. H. (1985). Managing new information technology: Design or default? *Human Resource Management Trends and Challenges* (69-102). Boston: Harvard Business School Press.

Phillips, L. (1983). A theoretical perspective on heuristics and biases in probabilistic thinking. In Humphreys, P., Swenson, O., & Vari, A. (Eds.), *Analysing and aiding decision processes* (27-43). Amsterdam: North Holland.

Phillips, L. (1985). Decision support for managers. In Otway, H. & Peltu, M. (Eds.), *The managerial challenge of new office technology* (80-98). London: Butterworths.

Piore, M.J. & Sabel, C.F. (1984). *The second industrial divide*. New York: Basic Books.

Rice, R. E. (1984). *The new media: Communication, research, and technology*. Beverly Hills CA: Sage.

Richman, L. (1987, June 8). Software catches the team spirit. *Fortune*, 125-133.

Rockart, J., & Treacy, M. (1981). *Executive information support systems (Technical report 65)*. Cambridge, MA: Center for Information Systems Research, Massachusetts Institute of Technology.

Rockart, J., & Treacy, M. (1982, January-February). The CEO goes on-line. *Harvard Business Review, 60*, 82-88.

Rousseau, D. M. (1983). Technology in organizations: A constructive review and analytic framework. In Seashore, S. E., Lawler, E. E., Mirvis, P. H., & Camman, C. (Eds.), *Assessing organizational changes: A guide to methods, measures and practices*. New York: Wiley & Sons.

Saffo, P. (1987, May). Invasion of the laser crud. *Personal Computing, 57*.

Salton, G. & McGill, M. (1983). *Introduction to modern information retrieval*. New York: McGraw-Hill.

Sarin, S., & Greif, I. (1985, October). Computer-based real-time conferencing systems. *IEEE Computer, 18*, 33-49.

Sathi, A., Fox, M., & Greenberg, M. (1985). Representation of activity knowledge for project management. *IEEE Transactions on Pattern Analysis and Machine Intelligence PAMI-7, 5*, 531-552.

Scott Morton, M. (1983). *State of the art of research in management support systems (Technical report 107)*. Cambridge, MA: Center for Information Systems Research, Massachusetts Institute of Technology.

Semprevivo, P. (1980). *Teams in information systems development*. New York: Yourdon Inc.

Simon, H. (1969). *The sciences of the artificial*. Cambridge, MA: MIT Press.

Sluizer, S., & Cashman, P. (1985). XCP: An experimental tool for supporting office procedures. *Proceedings of the Office Automation Conference* (73-80). Los Alamitos, CA: Institute of Electrical and Electronics Engineers.

Stasz, C. & Bikson, T. K. (1986). Computer-supported cooperative work: Examples and issues in one federal agency. *Proceedings of the Conference on*

Computer-Supported Cooperative Work (318-324). Austin TX: Microelectronics and Computer Technology Corp.

Stasz, C., Bikson, T. K.,& Shapiro, N. Z. (1986). *Assessing the Forest Service's implementation of an agency-wide information system: An exploratory study (Technical Report N-2463-USFS00)*. Santa Monica, CA: The RAND Corporation.

Stefik, M. (1986a). "WYSIWIS" revisited: Early experiences with multi-user interfaces. *Proceedings of the Conference on Computer-Supported Cooperative Work* (276-290). Austin, TX: Microelectronics and Computer Technology Corp.

Stefik, M. (1986b, Spring). The next knowledge medium. *AI Magazine, 7*, 34-46.

Stefik, M., Foster, G., Bobrow, D., Kahn, K., Lanning, S., & Suchman, L. (1987). Beyond the chalkboard: Using computers to support collaboration and problem solving in meetings. *Communications of the ACM, 30(1)*, 32-47.

Stroll, D., & Miller, L. (1984). *The executive research project (Technical report)*. Hudson, MA: Digital Equipment Corporation.

Suchman, L., & Trigg, R. (1986). A framework for studying research collaboration. *Proceedings of the Conference on Computer-Supported Cooperative Work* (221-228). Austin, TX: Microelectronics and Computer Technology Corp.

Systems Concepts Laboratory. (1987). *The office design project (A videotape)*. Palo Alto, CA: Xerox Palo Alto Research Center.

Talbert, L. R., Bikson, T. K., & Shapiro, N. Z. (1984). *Interactive information environments: A plan for enabling interdisciplinary research (Technical Report N-2115)*. Santa Monica, CA: The RAND Corporation.

Taylor, J. C. (1987). Job design and quality of working life. In Kraut, R. (Ed.), *Technology and the transformation of white collar work* (211-236). Hillsdale NJ: Lawrence Erlbaum Associates.

Thompson, G., (1975). An assessment methodology for evaluating communications innovations. *IEEE Transactions on Communications, COM-23*, 10, 1048.

Tornatzky, L. G. (1983). *The process of technological innovation: Reviewing the literature*. Washington, DC: National Science Foundation.

Tou, F. N., Williams, M. D., Fikes, R. E., Henderson, D. A., & Malone, T. W. (1982). RABBIT: An intelligent database assistant. *Proceedings of the National Conference of the American Association for Artificial Intelligence*, Pittsburgh, Pennsylvania.

Trigg, R., Suchman, L., & Halasz, F. (1986). Supporting collaboration with Note-Cards. *Proceedings of the Conference on Computer-Supported Cooperative Work* (153-162). Austin, TX: Microelectronics and Computer Technology Corp.

Trist, E. L. (1981). The sociotechnical perspective. In Van de Ven, A. H. & Joyce, W. F. (Eds.), *Perspectives on organization, design and behavior*. New York: John Wiley & Sons.

Wharton, A. (1987). *Report to Lotus Development (UK) and the Institute of Directors*. London: Wharton Information Systems.

Williamson, O.E. (1975). *Markets and hierarchies*. New York: Free Press.

Wilson, P., Maude, T., Marshall, C., & Heaton, N. (1984). *The active mailbox - Your on-line secretary*. Paper presented to IFIP Conference on Computer Based Message Services, Nottingham University.

Winograd, T. (1986). A language perspective on the design of cooperative work, *Proceedings of the Conference on Computer-Supported Cooperative Work*, (203-220). Austin, TX: Microelectronics and Computer Technology Corp.

Winograd, T., & Flores, F. (1986). *Understanding computers and cognition: A new foundation for design*. Norwood, NJ: Ablex Press.

Zisman, M. (1977). *Representation, specification, and implementation of office procedures*. Philadelphia, PA: Wharton School of Business, University of Pennsylvania.

AUTHOR INDEX

Subject Index